Expressions and Formulas

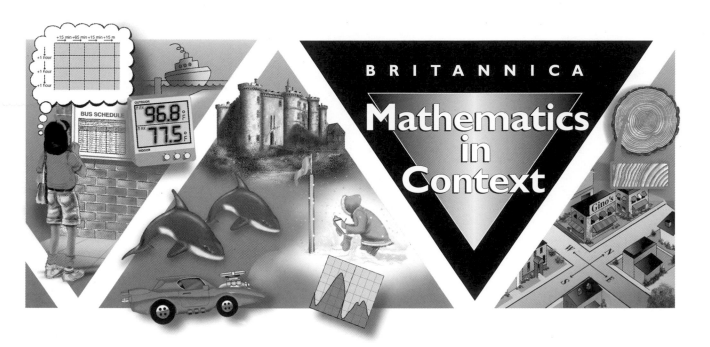

BRITANNICA

Mathematics in Context

Teacher Guide

Britannica

Mathematics in Context is a comprehensive curriculum for the middle grades. It was developed in collaboration with the Wisconsin Center for Education Research, School of Education, University of Wisconsin–Madison and the Freudenthal Institute at the University of Utrecht, The Netherlands, with the support of National Science Foundation Grant No. 9054928.

National Science Foundation

Opinions expressed are those of the authors
and not necessarily those of the Foundation

Revision Project

Peter Sickler
Project Director

Teri Hedges
Revision Consultant

Nieka Mamczak
Revision Consultant

Erin Turner
Revision Consultant

Cheryl Deese
MiC General Manager

Vicki Mirabile
Project Manager

The *Mathematics in Context* Development Team

Mathematics in Context is a comprehensive curriculum for the middle grades. The National Science Foundation funded the National Center for Research in Mathematical Sciences Education at the University of Wisconsin–Madison to develop and field-test the materials from 1991 through 1996. The Freudenthal Institute at the University of Utrecht in The Netherlands, as a subcontractor, collaborated with the University of Wisconsin–Madison on the development of the curriculum.

The initial version of *Expressions and Formulas* was developed by Koeno Gravemeijer, Anton Roodhardt, and Monica Wijers. It was adapted for use in American schools by Beth R. Cole and Gail Burrill.

National Center for Research in Mathematical Sciences Education Staff

Thomas A. Romberg
Director

Joan Daniels Pedro
Assistant to the Director

Gail Burrill
Coordinator
Field Test Materials

Margaret R. Meyer
Coordinator
Pilot Test Materials

Mary Ann Fix
Editorial Coordinator

Sherian Foster
Editorial Coordinator

James A. Middleton
Pilot Test Coordinator

Margaret A. Pligge
First Edition Coordinator

Project Staff

Jonathan Brendefur
Laura J. Brinker
James Browne
Jack Burrill
Rose Byrd
Peter Christiansen
Barbara Clarke
Doug Clarke
Beth R. Cole

Fae Dremock
Jasmina Milinkovic
Mary C. Shafer
Julia A. Shew
Aaron N. Simon
Marvin Smith
Stephanie Z. Smith
Mary S. Spence

Freudenthal Institute Staff

Jan de Lange
Director

Els Feijs
Coordinator

Martin van Reeuwijk
Coordinator

Project Staff

Mieke Abels
Nina Boswinkel
Frans van Galen
Koeno Gravemeijer
Marja van den Heuvel-Panhuizen
Jan Auke de Jong
Vincent Jonker
Ronald Keijzer

Martin Kindt
Jansie Niehaus
Nanda Querelle
Anton Roodhardt
Leen Streefland
Adri Treffers
Monica Wijers
Astrid de Wild

Acknowledgments

Several school districts used and evaluated one or more versions of the materials: Ames Community School District, Ames, Iowa; Parkway School District, Chesterfield, Missouri; Stoughton Area School District, Stoughton, Wisconsin; Madison Metropolitan School District, Madison, Wisconsin; Milwaukee Public Schools, Milwaukee, Wisconsin; and Dodgeville School District, Dodgeville, Wisconsin. Two sites were involved in staff developments as well as formative evaluation of materials: Culver City, California, and Memphis, Tennessee. Two sites were developed through partnership with Encyclopædia Britannica, Inc.: Miami, Florida, and Puerto Rico. University Partnerships were developed with mathematics educators who worked with preservice teachers to familiarize them with the curriculum and to obtain their advice on the curriculum materials. The materials were also used at several other schools throughout the United States.

We at Encyclopædia Britannica, Inc. extend our thanks to all who had a part in making this program a success. Some of the participants instrumental in the program's development are as follows:

Allapattah Middle School
Miami, Florida
Nemtalla (Nikolai) Barakat

Ames Middle School
Ames, Iowa
Kathleen Coe
Judd Freeman
Gary W. Schnieder
Ronald H. Stromen
Lyn Terrill

Bellerive Elementary
Creve Coeur, Missouri
Judy Hetterscheidt
Donna Lohman
Gary Alan Nunn
Jakke Tchang

Brookline Public Schools
Brookline, Massachusetts
Rhonda K. Weinstein
Deborah Winkler

Cass Middle School
Milwaukee, Wisconsin
Tami Molenda
Kyle F. Witty

Central Middle School
Waukesha, Wisconsin
Nancy Reese

Craigmont Middle School
Memphis, Tennessee
Sharon G. Ritz
Mardest K. VanHooks

Crestwood Elementary
Madison, Wisconsin
Diane Hein
John Kalson

Culver City Middle School
Culver City, California
Marilyn Culbertson
Joel Evans
Joy Ellen Kitzmiller
Patricia R. O'Connor
Myrna Ann Perks, Ph.D.
David H. Sanchez
John Tobias
Kelley Wilcox

Cutler Ridge Middle School
Miami, Florida
Lorraine A. Valladares

Dodgeville Middle School
Dodgeville, Wisconsin
Jacqueline A. Kamps
Carol Wolf

Edwards Elementary
Ames, Iowa
Diana Schmidt

Fox Prairie Elementary
Stoughton, Wisconsin
Tony Hjelle

Grahamwood Elementary
Memphis, Tennessee
M. Lynn McGoff
Alberta Sullivan

Henry M. Flagler Elementary
Miami, Florida
Frances R. Harmon

Horning Middle School
Waukesha, Wisconsin
Connie J. Marose
Thomas F. Clark

Huegel Elementary
Madison, Wisconsin
Nancy Brill
Teri Hedges
Carol Murphy

Hutchison Middle School
Memphis, Tennessee
Maria M. Burke
Vicki Fisher
Nancy D. Robinson

Idlewild Elementary
Memphis, Tennessee
Linda Eller

Jefferson Elementary
Santa Ana, California
Lydia Romero-Cruz

Jefferson Middle School
Madison, Wisconsin
Jane A. Beebe
Catherine Buege
Linda Grimmer
John Grueneberg
Nancy Howard
Annette Porter
Stephen H. Sprague
Dan Takkunen
Michael J. Vena

Jesus Sanabria Cruz School
Yabucoa, Puerto Rico
Andreíta Santiago Serrano

John Muir Elementary School
Madison, Wisconsin
Julie D'Onofrio
Jane M. Allen-Jauch
Kent Wells

Kegonsa Elementary
Stoughton, Wisconsin
Mary Buchholz
Louisa Havlik
Joan Olsen
Dominic Weisse

Linwood Howe Elementary
Culver City, California
Sandra Checel
Ellen Thireos

Mitchell Elementary
Ames, Iowa
Henry Gray
Matt Ludwig

New School of Northern Virginia
Fairfax, Virginia
Denise Jones

Northwood Elementary
Ames, Iowa
Eleanor M. Thomas

Orchard Ridge Elementary
Madison, Wisconsin
Mary Paquette
Carrie Valentine

Parkway West Middle School
Chesterfield, Missouri
Elissa Aiken
Ann Brenner
Gail R. Smith

Ridgeway Elementary
Ridgeway, Wisconsin
Lois Powell
Florence M. Wasley

Roosevelt Elementary
Ames, Iowa
Linda A. Carver

Roosevelt Middle
Milwaukee, Wisconsin
Sandra Simmons

Ross Elementary
Creve Coeur, Missouri
Annette Isselhard
Sheldon B. Korklan
Victoria Linn
Kathy Stamer

St. Joseph's School
Dodgeville, Wisconsin
Rita Van Dyck
Sharon Wimer

St. Maarten Academy
St. Peters, St. Maarten, NA
Shareed Hussain

Sarah Scott Middle School
Milwaukee, Wisconsin
Kevin Haddon

Sawyer Elementary
Ames, Iowa
Karen Bush Hoiberg

Sennett Middle School
Madison, Wisconsin
Brenda Abitz
Lois Bell
Shawn M. Jacobs

Sholes Middle School
Milwaukee, Wisconsin
Chris Gardner
Ken Haddon

Stephens Elementary
Madison, Wisconsin
Katherine Hogan
Shirley M. Steinbach
Kathleen H. Vegter

Stoughton Middle School
Stoughton, Wisconsin
Sally Bertelson
Polly Goepfert
Jacqueline M. Harris
Penny Vodak

Toki Middle School
Madison, Wisconsin
Gail J. Anderson
Vicky Grice
Mary M. Ihlenfeldt
Steve Jernegan
Jim Leidel
Theresa Loehr
Maryann Stephenson
Barbara Takkunen
Carol Welsch

Trowbridge Elementary
Milwaukee, Wisconsin
Jacqueline A. Nowak

W. R. Thomas Middle School
Miami, Florida
Michael Paloger

Wooddale Elementary Middle School
Memphis, Tennessee
Velma Quinn Hodges
Jacqueline Marie Hunt

Yahara Elementary
Stoughton, Wisconsin
Mary Bennett
Kevin Wright

Site Coordinators

Mary L. Delagardelle—Ames Community Schools, Ames, Iowa

Dr. Hector Hirigoyen—Miami, Florida

Audrey Jackson—Parkway School District, Chesterfield, Missouri

Jorge M. López—Puerto Rico

Susan Militello—Memphis, Tennessee

Carol Pudlin—Culver City, California

Reviewers and Consultants

Michael N. Bleicher
Professor of Mathematics
University of Wisconsin–Madison
Madison, WI

Diane J. Briars
Mathematics Specialist
Pittsburgh Public Schools
Pittsburgh, PA

Donald Chambers
Director of Dissemination
University of Wisconsin–Madison
Madison, WI

Don W. Collins
Assistant Professor of Mathematics Education
Western Kentucky University
Bowling Green, KY

Joan Elder
Mathematics Consultant
Los Angeles Unified School District
Los Angeles, CA

Elizabeth Fennema
Professor of Curriculum and Instruction
University of Wisconsin–Madison
Madison, WI

Nancy N. Gates
University of Memphis
Memphis, TN

Jane Donnelly Gawronski
Superintendent
Escondido Union High School
Escondido, CA

M. Elizabeth Graue
Assistant Professor of Curriculum and Instruction
University of Wisconsin–Madison
Madison, WI

Jodean E. Grunow
Consultant
Wisconsin Department of Public Instruction
Madison, WI

John G. Harvey
Professor of Mathematics and Curriculum & Instruction
University of Wisconsin–Madison
Madison, WI

Simon Hellerstein
Professor of Mathematics
University of Wisconsin–Madison
Madison, WI

Elaine J. Hutchinson
Senior Lecturer
University of Wisconsin–Stevens Point
Stevens Point, WI

Richard A. Johnson
Professor of Statistics
University of Wisconsin–Madison
Madison, WI

James J. Kaput
Professor of Mathematics
University of Massachusetts–Dartmouth
Dartmouth, MA

Richard Lehrer
Professor of Educational Psychology
University of Wisconsin–Madison
Madison, WI

Richard Lesh
Professor of Mathematics
University of Massachusetts–Dartmouth
Dartmouth, MA

Mary M. Lindquist
Callaway Professor of Mathematics Education
Columbus College
Columbus, GA

Baudilio (Bob) Mora
Coordinator of Mathematics & Instructional Technology
Carrollton-Farmers Branch Independent School District
Carrollton, TX

Paul Trafton
Professor of Mathematics
University of Northern Iowa
Cedar Falls, IA

Norman L. Webb
Research Scientist
University of Wisconsin–Madison
Madison, WI

Paul H. Williams
Professor of Plant Pathology
University of Wisconsin–Madison
Madison, WI

Linda Dager Wilson
Assistant Professor
University of Delaware
Newark, DE

Robert L. Wilson
Professor of Mathematics
University of Wisconsin–Madison
Madison, WI

TABLE OF CONTENTS

BRITANNICA
Mathematics
in
Context

Dear Teacher,

Welcome! *Mathematics in Context* is designed to reflect the National Council of Teachers of Mathematics Standards for School Mathematics and to ground mathematical content in a variety of real-world contexts. Rather than relying on you to explain and demonstrate generalized definitions, rules, or algorithms, students investigate questions directly related to a particular context and construct mathematical understanding and meaning from that context.

The curriculum encompasses 10 units per grade level. This unit is designed to be the first unit in the algebra strand for grade 6/7, but it also lends itself to independent use—to describe and express complex ideas and generalities.

In addition to the Teacher Guide and Student Books, *Mathematics in Context* offers the following components that will inform and support your teaching:

- *Teacher Resource and Implementation Guide*, which provides an overview of the complete system, including program implementation, philosophy, and rationale

- *Number Tools*, which is a series of blackline masters that serve as review sheets or practice pages involving number issues and basic skills

- *Mathematics in Context Online*, which is a rich, balanced resource for teachers, students, and parents looking for additional information, activities, tools, and support to further students' mathematical understanding and achievements

- *News in Numbers*, which is a volume of additional activities that can be inserted between or within other units

Thank you for choosing *Mathematics in Context.* We wish you success and inspiration!

Sincerely,

The Mathematics in Context Development Team

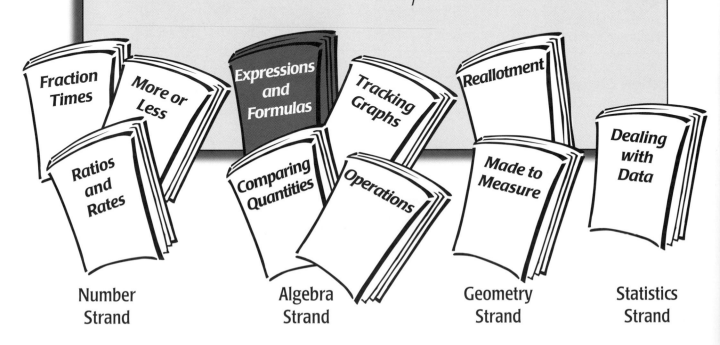

Number Strand   Algebra Strand   Geometry Strand   Statistics Strand

Overview

BRITANNICA

Mathematics in Context

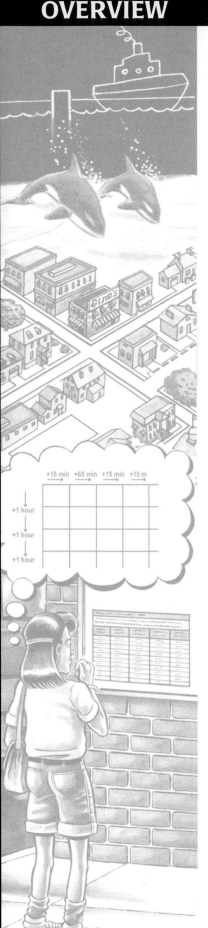

How to Use This Book

This unit is one of 40 for the middle grades. Each unit can be used independently; however, the 40 units are designed to make up a complete, connected curriculum (10 units per level). There is a Student Book and a Teacher Guide for each unit.

Each Teacher Guide comprises elements that assist the teacher in the presentation of concepts and in understanding the general direction of the unit and the program as a whole. Becoming familiar with this structure will make using the units easier.

Each Teacher Guide consists of six basic parts:

- Overview
- Student Materials and Teaching Notes
- Assessment Activities and Solutions
- Glossary
- Blackline Masters
- Try This! Solutions

Overview

Before beginning this unit, read the Overview in order to understand the purpose of the unit and to develop strategies for facilitating instruction. The Overview provides helpful information about the unit's focus, materials preparation, pacing and planning, goals and assessment, as well as explanations about how the unit fits with the rest of the *Mathematics in Context* curriculum.

Student Materials and Teaching Notes

This Teacher Guide contains all of the student pages, each of which faces a page of solutions, samples of students' work, and hints and comments about how to facilitate instruction.

SECTION D. REVERSE OPERATIONS

Section Focus

Students use reverse arrow strings to solve problems involving exchange rates and other contexts. The instructional focus of Section D is to:
- Use reverse arrow strings to solve problems.
- Use arrow string formulas to represent calculations with decimals and ratios.

Planning Instruction

Day 10. Foreign Money   Student pages 27–29

INTRODUCTION	Problems 1 and 2	■ Use an exchange rate to convert U.S. dollars into Dutch guilders.
CLASSWORK	Problems 3–10	■ Use arrow string formulas to convert U.S. dollars into Dutch guilders and vice versa.
HOMEWORK	Problems 11 and 12	■ Convert U.S. prices in dollars into Dutch guilders.

Additional Resources: Extension, page 71

Day 11. Going Backwards   Student pages 30–32

INTRODUCTION	Problems 13 and 14	■ Discuss a game involving reverse arrow strings.
CLASSWORK	Problems 15–19	■ Use reverse arrow strings to calculate the age of a tree.
HOMEWORK	Problems 20–22	■ Use arrow strings to determine the cost of deli items.

Additional Resources: Extension, page 73; Writing Opportunity, page 77; Number Tools, Volume 2, page 12

Day 12. Summary   Student page 32

| INTRODUCTION | Review homework | ■ Review homework from Day 11. |
| ASSESSMENT | Problems 23 and 24 | ■ Summary Questions. |

Additional Resources: Try This! Section D, Student page 57; Number Tools, Volume 2, page 14

SECTION D. REVERSE OPERATIONS

Materials

| Student Resources | Teacher Resources | Student Materials |
| No resources required. | No resources required. | No materials required. |

*See Hints and Comments for optional materials.

Concept Development

Reverse Arrow Strings to Solve Problems

Section D introduces the use of reverse arrow strings to solve problems involving exchange rates.

On **Day 10**, students investigate an exchange rate to convert U.S. dollars into Dutch guilders and vice versa. They use arrow string formulas that involve multiplication and division to represent and perform calculations with decimals and ratios.

On **Day 11**, students discuss a game involving reverse arrow strings. Given the arrow string and the output (the answer), they identify a strategy for finding the input (the starting number) of the formula. Next, students use reverse arrow strings to calculate the age of a tree. For homework, students use arrow strings to determine the cost of deli items.

On **Day 12**, students review homework from Day 11 and demonstrate their ability to reverse arrow strings and their understanding of the usefulness of reverse strings.

Planning Assessment

Problem 7
- Use and interpret simple formulas.
- Use reverse operations to find the input for a given output.
- Solve problems using the relationship between a mathematical procedure and its inverse.

Problem 14
- Use reverse operations to find the input for a given output.
- Solve problems using the relationship between a mathematical procedure and its inverse.

64   Section D Reverse Operations   Britannica Mathematics System

Mathematics in Context • Expressions and Formulas   Section D Reverse Operations 65

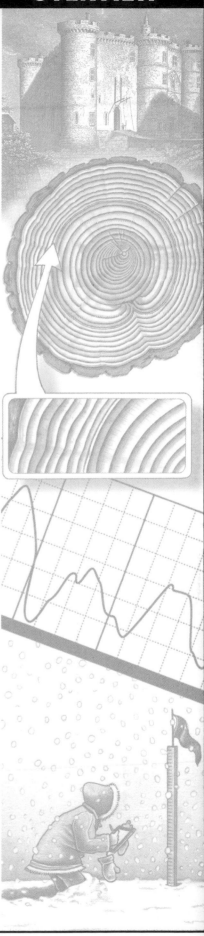

Each section within the unit begins with a two-page overview that describes the instructional focus of the section, provides day–by–day outline of lessons, details the development of concepts, lists the necessary materials and resources, and highlights the ongoing assessment of unit goals.

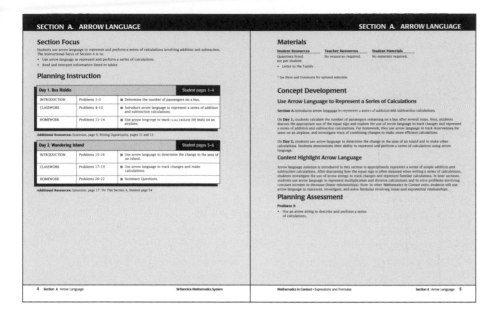

Assessment Activities and Solutions

Information about assessment can be found in several places in this Teacher Guide. General information about assessment is given in the Overview; informal assessment opportunities are identified on the teacher pages that face each student page; and the Assessment Activities section of this guide provides formal assessment opportunities.

Glossary

The Glossary defines all vocabulary words listed on the Section Opener pages. It includes the mathematical terms that may be new to students, as well as words associated with the contexts introduced in the unit. (Note: The Student Book does not have a glossary. This allows students to construct their own definitions, based on their personal experiences with the unit activities.)

Blackline Masters

At the back of this Teacher Guide are blackline masters for photocopying. The blackline masters include a letter to families (to be sent home with students before beginning the unit), several student activity sheets, and assessment masters.

Try This! Solutions

Also included in the back of this Teacher Guide are the solutions to several *Try This!* activities—one related to each section of the unit—that can be used to reinforce the unit's main concepts. The *Try This!* activities from the Student Book are also shown.

78%

Unit Focus

Students learn to use *arrow language* to accurately represent a series of calculations. After using these arrow strings to simplify calculations, students write, evaluate, and solve arrow string formulas to solve problems involving constant increase and rates (linear relationships). In addition, students learn the order of operations. The instructional focus of this unit is to:

• Use *arrow language* to represent and perform a series of calculations.

• Shorten and extend arrow strings to simplify calculations.

• Write and evaluate arrow string formulas to solve problems.

• Use reverse arrow strings to solve problems.

• Use order of operations to perform a series of calculations.

Prior Knowledge

This unit assumes that students can perform computations with whole numbers, positive rational numbers, decimals, and fractions. Students should know how to add and subtract decimals, to multiply a decimal by a whole number by using repeated addition, and to use number sense approaches when calculating with decimals and fractions. Students should also be familiar with using a ratio table.

This unit should be taught after the algebra units *Patterns and Symbols* and *Dry and Wet Numbers* and the number units *Some of the Parts*, *Per Sense*, and *Measure for Measure*.

Materials Preparation

The following items are the necessary materials and resources to be used by the teacher and students throughout the unit. For further details, see the Section Overviews and the Materials part of the Hints and Comments section at the top of each teacher page. Note: Some contexts and problems can be enhanced through the use of optional materials. These optional materials are listed in the corresponding Hints and Comments section.

Student Resources

Quantities listed are per student.

• Letter to the Family
• Student Activity Sheets 1–5

Teacher Resources

No resources required.

Student Materials

Quantities listed are per student, unless otherwise noted.

• Calculator (one per student)
• Centimeter ruler
• Styrofoam® cups, four

Pacing and Planning
Pacing: 17 days

Section A. Arrow Language	2 Days (in total)

Students use arrow language to represent and perform a series of calculations involving addition and subtraction.

| Day 1 Bus Riddle | Introduce arrow language to represent a series of addition and subtraction calculations. |
| Day 2 Wandering Island | Use arrow language to track changes and make calculations. |

Section B. Smart Calculations	2 Days (in total)

Students investigate strategies for determining the correct change from a purchase and for rewriting numerical expressions to simplify calculations.

| Day 3 Making Change | Use arrow strings to calculate correct change from a purchase. |
| Day 4 Skillful Computations (Continued) | Shorten and extend arrow strings to simplify calculations. |

Section C. Formulas	5 Days (in total)

Students use arrow strings to represent multiplication and division calculations and to write arrow string formulas to represent and solve problems.

Day 5 Supermarket	Use arrow strings that involve multiplication to calculate the cost of produce.
Day 6 Taxi Fares	Use arrow strings to compare different rates for taxicab rides.
Day 7 Stacking Cups	Determine the height of a stack of cups using arrow string formulas.
Day 8 Bike Sizes	Write an arrow string to represent a formula for determining the dimensions of a bike.
Day 9 Summary	Write, evaluate, and solve arrow string formulas to solve problems.

Section D. Reverse Operations — 3 Days (in total)

Students use reverse arrow strings to solve problems involving exchange rates and other contexts.

Day 10 Foreign Money	Use arrow string formulas to convert U.S. dollars into Dutch guilders and vice versa.
Day 11 Going Backwards	Use reverse arrow strings to calculate the age of a tree.
Day 12 Summary	Write reverse arrow strings.

Section E. Tables — 2 Days (in total)

Students use reverse arrow strings to solve problems involving exchange rates and other contexts.

Day 13 Apple Crisp	Use arrow strings to describe the numbers in a bus schedule table.
Day 14 Home Repairs	Calculate the cost of plumbing repairs.

Section F. Order of Operations — 3 Days (in total)

Students use number sentences, arrow strings, and arithmetic trees to investigate the order of operations for a series of calculations.

Day 15 Arithmetic Trees	Use order of operations and arithmetic trees to perform a series of calculations.
Day 16 Flexible Computation	Use arithmetic trees to make flexible calculations.
Day 17 Return to the Supermarket (Continued)	Use parentheses to replace arithmetic trees in calculations.

45%

Assessment

Planning Assessment

In keeping with the NCTM Assessment Standards, valid assessment should be based on evidence drawn from several sources. (See the full discussion of assessment philosophies in the *Teacher Resource and Implementation Guide*.) An assessment plan for this unit may draw from the following sources:

- Observations—look, listen, and record observable behavior.

- Interactive Responses—in a teacher-facilitated situation, note how students respond, clarify, revise, and extend their thinking.

- Products—look for the quality of thought evident in student projects, test answers, worksheet solutions, or writings.

These categories are not meant to be mutually exclusive. In fact, observation is a key part in assessing interactive responses and also a key to understanding the end results of projects and writings.

Ongoing Assessment Opportunities

- **Problems within Sections**
  To evaluate ongoing progress, *Mathematics in Context* identifies informal assessment opportunities and the goals that these particular problems assess throughout the Teacher Guide. There are also indications as to what you might expect from your students.

- **Section Summary Questions**
  The summary questions at the end of each section are vehicles for informal assessment (see Teacher Guide pages 16, 34, 60, 62, 76, 94, and 122).

End-of-Unit Assessment Opportunities

In the back of this Teacher Guide, there are three assessment activities that can be completed in one 45-minute class period. For a more detailed description of the assessment activities, see the Assessment Overview (Teacher Guide pages 124 and 125).

You may also wish to design your own culminating project or let students create one that will tell you what they consider is important in the unit. For more assessment ideas, refer to the chart on pages xvii and xviii.

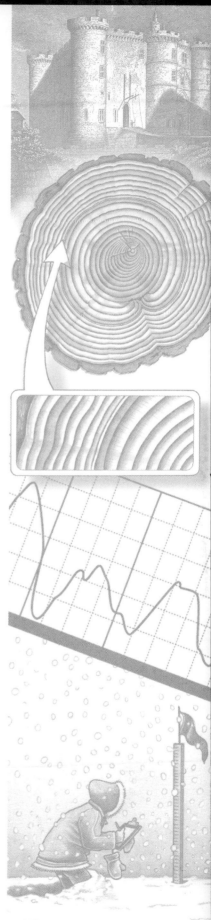

More about Assessment

Scoring and Analyzing Assessment Responses

Students may respond to assessment questions with various levels of mathematical sophistication and elaboration. Each student's response should be considered for the mathematics that it shows, and not judged on whether or not it includes an expected response. Responses to some of the assessment questions may be viewed as either correct or incorrect, but many answers will need flexible judgment by the teacher. Descriptive judgments related to specific goals and partial credit often provide more helpful feedback than percentage scores.

Openly communicate your expectations to all students, and report achievement and progress for each student relative to those expectations. When scoring students' responses, try to think about how they are progressing toward the goals of the unit and the strand.

Student Portfolios

Generally, a portfolio is a collection of student-selected pieces that is representative of a student's work. A portfolio may include evaluative comments by you or by the student. See the *Teacher Resource and Implementation Guide* for more ideas on portfolio focus and use.

A comprehensive discussion about the contents, management, and evaluation of portfolios can be found in *Mathematics Assessment: Myths, Models, Good Questions, and Practical Suggestions,* pp. 35–48.

Student Self-Evaluation

Self-evaluation encourages students to reflect on their progress in learning mathematical concepts, their developing abilities to use mathematics, and their dispositions toward mathematics. The following examples illustrate ways to incorporate student self-evaluations as one component of your assessment plan.

- Ask students to comment, in writing, on each piece they have chosen for their portfolios and on the progress they see in the pieces overall.
- Give a writing assignment entitled "What I Know Now about [a math concept] and What I Think about It." This will give you information about each student's disposition toward mathematics as well as his or her knowledge.
- Interview individuals or small groups to elicit what they have learned, what they think is important, and why.

Suggestions for self-inventories can be found in *Mathematics Assessment: Myths, Models, Good Questions, and Practical Suggestions,* pp. 55–58.

Summary Discussion

Discuss specific lessons and activities in the unit—what the student learned from them and what the activities have in common. This can be done in whole-class discussion, in small groups, or in personal interviews.

Goals and Assessment

In the *Mathematics in Context* curriculum, unit goals, categorized according to cognitive procedures, relate to the strand goals and the *NCTM Principles and Standards for School Mathematics (Standards 2000 Project)*. Additional information about these goals is found in the *Teacher Resource and Implementation Guide*. The *Mathematics in Context* curriculum is designed to help students develop their abilities so that they can perform with understanding in each of the categories listed below. It is important to note that the attainment of goals in one category is not a prerequisite to attaining those in another category. In fact, students should progress simultaneously toward several goals in different categories.

	Goal	Ongoing Assessment Opportunities	End-of-Unit Assessment Opportunities
Conceptual and Procedural Knowledge	**1.** describe and perform a series of calculations using an arrow string	**Section A** p. 10, #9 **Section B** p. 26, #7 p. 34, #18	Get Your Programs Here!, p. 142 Forest Fire Fighting, pp. 143 and 144 Expressions, p. 145
	2. describe and perform a series of calculations using an arithmetic tree	**Section F** p. 112, #16, #19 p. 114, #24 p. 116, #25 p. 122, #31	Get Your Programs Here!, p. 142 Expressions, p. 145
	3. use and interpret simple formulas	**Section C** p. 46, #10 p. 52, #23, #24 p. 62, #39 **Section D** p. 70, #7 **Section E** p. 92, #16 **Section F** p. 112, #16 p. 116, #25 p. 122, #31	Get Your Programs Here!, p. 142 Forest Fire Fighting, pp. 143 and 144 Expressions, p. 145
	4. use conventional rules and grouping symbols to perform a sequence of calculations	**Section F** p. 122, #31	Expressions, p. 145
	5. use reverse operations to find the input for a given output	**Section C** p. 46, #11 p. 52, #24 Assessment Opportunity, p. 53 p. 62, #38 **Section D** p. 70, #7 p. 72, #14	Get Your Programs Here!, p. 142 Forest Fire Fighting, pp. 143 and 144

	Goal	Ongoing Assessment Opportunities	End-of-Unit Assessment Opportunities
Reasoning, Communicating, Thinking, and Making Connections	**6.** rewrite numerical expressions to facilitate calculation	**Section B** p. 34, #16, #17 Assessment Opportunity, p. 31 **Section C** p. 48, #16, #17 **Section F** p. 112, #16	Expressions, p. 145
	7. reason from a series of calculations to an informal formula	**Section C** p. 52, #23, #24 p. 62, #37 **Section F** p. 122, #31	Get Your Programs Here!, p. 142 Expressions, p. 145
	8. interpret relationships displayed in tables	**Section E** p. 86, #9 p. 88, #11 p. 92, #15, #16	Get Your Programs Here!, p. 142 Expressions, p. 145
	9. use word variables to describe a formula or procedure	**Section C** p. 52, #23, #24 **Section F** p. 116, #25	Forest Fire Fighting, pp. 143 and 144

	Goal	Ongoing Assessment Opportunities	End-of-Unit Assessment Opportunities
Modeling, Nonroutine Problem-Solving, Critically Analyzing, and Generalizing	**10.** generalize from patterns to symbolic relationships	**Section E** p. 92, #16	Get Your Programs Here!, p. 142 Forest Fire Fighting, pp. 143 and 144 Expressions, p. 145
	11. solve problems using the relationship between a mathematical procedure and its inverse	**Section D** p. 70, #7 p. 72, #14	Get Your Programs Here!, p. 142 Forest Fire Fighting, pp. 143 and 144
	12. use formulas in any representation (arrow language, arithmetic tree, words) to solve problems	**Section C** Assessment Opportunity, p. 53 p. 62, #36 **Section E** p. 92, #16 **Section F** p. 112, #19 p. 114, #24 p. 116, #25	Get Your Programs Here!, p. 142 Forest Fire Fighting, pp. 143 and 144

Connections with Other *Mathematics in Context* Units

Expressions and Formulas has strong connections with the number strand units. Students learn smart ways to calculate, by breaking up a calculation into easier calculations or by reorganizing the calculation.

The ratio table model, introduced in *Some of the Parts* and further developed in *Per Sense,* is also used in *Expressions and Formulas.*

The rules for the orders of operations are introduced in this unit and studied more formally in the number unit *Reflections on Number.* Arrow language is also revisited in the number unit *More or Less.* In *Expressions and Formulas,* students are not expected to know formal algorithms for calculating with rational numbers. More complex operations with rational numbers are developed in the number units *More or Less* and *Ratio and Rates.* Connections between fractions and decimals are made in *Expressions and Formulas* and made more explicit in the unit *Fraction Times.* The work with tables and charts in *Expressions and Formulas* is frequently revisited in units in the statistics strand, such as *Ways to Go.*

The following mathematical topics that are included in the unit *Expressions and Formulas* are introduced or further developed in other *Mathematics in Context* units.

Prerequisite Topics

Topic	Unit	Grade
ratio table	*Some of the Parts* **	5/6
	Per Sense **	5/6
money	*Measure for Measure* **	5/6
fractions and decimals	*Fraction Times* **	6/7
patterns	*Patterns and Symbols*	5/6

Topics Revisited in Other Units

Topic	Unit	Grade
patterns	*Comparing Quantities*	6/7
	Building Formulas	7/8
formulas	*Building Formulas*	7/8
	Graphing Equations	8/9
	Patterns & Figures	8/9
	Digging Numbers *	8/9
	all algebra units	
order of operations	*Operations*	6/7
	Building Formulas	7/8
	Graphing Equations	8/9
calculation with rational numbers	*More or Less* **	6/7
	Ratios & Rates **	6/7
tables and charts	*Comparing Quantities*	6/7
	Ways to Go *	7/8
inverse operations	*Reflections on Number* **	8/9
arrow language	*More or Less* **	6/7

* These units in the statistics strand also help students make connections to ideas from the number strand.
** These units in the number strand also help students make connections to ideas about numbers.

Student
Materials
and Teaching
Notes

Student Book
Table of Contents

Dear Student,

Welcome to *Expressions and Formulas*.

Have you ever wondered how to figure out what size bike you need? There is a formula that uses the length of your legs to tell you. In this unit, you will see this and many other formulas. Sometimes you will figure out the formula by yourself. Sometimes the formula will be given to you, and you will use it to answer questions.

In this unit, you will also learn new forms of mathematical writing. You will use "arrow language," arithmetic trees, and parentheses. These different ways of writing will help you understand and use formulas and other math expressions.

As you work on this unit, you may want to keep an eye out for formulas that you see outside of math class. They are everywhere!

Sincerely,

The Mathematics in Context Development Team

Section Focus

Students use *arrow language* to represent and perform a series of calculations involving addition and subtraction. The instructional focus of Section A is to:

- Use arrow language to represent and perform a series of calculations.
- Read and interpret information listed in tables.

Planning Instruction

Day 1. Bus Riddle		Student pages 1–4
INTRODUCTION	Problems 1–3	■ Determine the number of passengers on a bus.
CLASSWORK	Problems 4–10	■ Introduce arrow language to represent a series of addition and subtraction calculations.
HOMEWORK	Problems 11–14	■ Use arrow language to track reservations for seats on an airplane.

Additional Resources: Extension, page 9; Writing Opportunity, pages 11 and 13

Day 2. Wandering Island		Student pages 5 and 6
INTRODUCTION	Problems 15 and 16	■ Use arrow language to determine the change in the area of an island.
CLASSWORK	Problems 17–19	■ Use arrow language to track changes and make calculations.
HOMEWORK	Problems 20–22	■ Summary Questions.

Additional Resources: Extension, page 17; Try This! Section A, Student page 54

Materials

Student Resources	**Teacher Resources**	**Student Materials**
Quantities listed are per student.	No resources required.	No materials required.
• Letter to the Family		

* See Hints and Comments for optional materials.

Concept Development

Use Arrow Language to Represent a Series of Calculations

Section A introduces *arrow language* to represent a series of addition and subtraction calculations.

On **Day 1,** students calculate the number of passengers remaining on a bus after several stops. Next, students discuss the appropriate use of the equal sign and explore the use of arrow language to track changes and represent a series of addition and subtraction calculations. For homework, they use arrow language to track reservations for seats on an airplane and investigate ways of combining changes to make more efficient calculations.

On **Day 2,** students use arrow language to determine the change in the area of an island and to make other calculations. Students demonstrate their ability to represent and perform a series of calculations using arrow language.

Content Highlight: Arrow Language

Arrow language notation is introduced in this section to appropriately represent a series of simple addition and subtraction calculations. After discussing how the equal sign is often misused when writing a series of calculations, students investigate the use of arrow strings to track changes and represent familiar calculations. In later sections, students use arrow language to represent multiplication and division calculations and to solve problems involving constant increase or decrease (linear relationships). Note: In other *Mathematics in Context* units, students will use arrow language to represent, investigate, and solve formulas involving linear and exponential relationships.

Planning Assessment

Problem 9
• Use an arrow string to describe and perform a series of calculations.

A. ARROW LANGUAGE

Bus Riddle

Pretend that you are a bus driver. Early one morning, you leave the garage with no passengers. At the first stop, 10 people get on the bus. At the next stop, six people get on. At the next stop, four people get off the bus, and seven get on. After that, five people get on, and two people get off. At the next stop, four people get off, and no one gets on.

1. How old is the bus driver?

2. Did you expect the first question to be about the number of passengers on the bus?

3. How could you determine the number of passengers on the bus after the last stop mentioned above?

1. Since the reader (or student) is pretending to be the bus driver, the answer is the reader's (or student's) current age.

2. Answers will vary, although most students will probably answer yes.

3. Answers and strategies will vary, but should include sequentially adding and subtracting the number of passengers getting on and off the bus.

Sample strategies:

Strategy 1

Some students may list the number of passengers getting on and off the bus at each stop.

stop 1   10

stop 2 $10 + 6 = 16$

stop 3 $16 - 4 + 7 = 19$

stop 4 $19 + 5 - 2 = 22$

stop 5 $22 - 4 = 18$

Strategy 2

Some students may list the activity of the passengers as a string of numerical operations.

$\underline{10 + 6} - 4 + 7 + 5 - 2 - 4$

$\quad \underline{16 - 4} + 7 + 5 - 2 - 4$

$\qquad \underline{12 + 7} + 5 - 2 - 4$

$\qquad\quad \underline{19 + 5} - 2 - 4$

$\qquad\qquad \underline{24 - 2} - 4$

$\qquad\qquad\quad \underline{22 - 4}$

$\qquad\qquad\qquad 18$

Overview Students read and solve a math riddle involving passengers getting on and off a bus.

About the Mathematics Many students read a problem and sometimes prepare to answer the wrong question. Problem **1** on this page will remind students to read the question before they try to answer it. Problem **3** gets students to think about how they might record a series of operations to solve a problem. This is an essential skill for this unit.

Comments about the Problems

1. Problem **1** is deliberately misleading. It has more to do with students' comprehension skills than with mathematics.

3. This problem illustrates the need for organizing and recording a series of calculations in one problem. Students may invent interesting ways to record the calculations. Watch for and correct students who use the equal sign incorrectly.

   For example, some students may use the equal sign to record the calculations for this problem in the following incorrect way:

   $10 + 6 = 16 - 4 + 7 = 19 + 5 - 2 = 22 - 4 = 18$

   This calculation is incorrect because it creates a string of untrue statements. The first statement, $10 + 6$, does not equal the second statement, $16 - 4 + 7$, and 16 does not equal 19 and so on. The calculation can be broken down into smaller parts as in Strategy 1 in the solutions column (in which the equal sign *is* used correctly).

   Using the equal sign correctly is addressed specifically in problem **6** on page 8. You may wait until then to discuss it with the entire class.

When four people got off the bus and seven got on, the number of people on the bus changed. There were three more people on the bus.

4. Here is a record of people getting on and off the bus for another part of the day. Copy the chart into your notebook. Then fill in the missing numbers.

Number of Passengers Getting off the Bus	Number of Passengers Getting on the Bus	Change
5	8	3 more
9	13	
16	16	
15	8	
9	3	
		5 fewer

5. Look at the last row in the chart. What can you say about the numbers of passengers getting on and off the bus when you know only that there are five fewer people on the bus?

For the story on page l, you might have kept track of the number of passengers on the bus by writing:

$$10 + 6 = 16 + 3 = 19 + 3 = 22 - 4 = 18$$

6. Do you think that writing the numbers in this way is acceptable in mathematics? Why or why not?

To avoid problems with the equals sign (=), you could write the calculation like this:

$$10 \xrightarrow{+6} 16 \xrightarrow{+3} 19 \xrightarrow{+3} 22 \xrightarrow{-4} 18$$

Each change is shown with an arrow.

This way of writing a string of calculations is called *arrow language.* You can use arrow language to describe any sequence of additions and subtractions, whether it is about passengers, money, or anything else.

7. Why is arrow language a better way to keep track of a changing total?

4.

Number of Passengers Getting off the Bus	Number of Passengers Getting on the Bus	Change
5	8	3 more
9	13	4 more
16	16	0
15	8	7 fewer
9	3	6 fewer
		5 fewer

5. Answers will vary. Any pair of numbers such that the number of passengers getting off the bus exceeds the number of passengers getting on the bus by 5 is acceptable. Possible answers include:

- 5 people getting off and 0 getting on
- 6 people getting off and 1 getting on
- 7 people getting off and 2 getting on

6. No. This notation is not acceptable because someone could read $10 + 6 = 16 + 3$ as $16 = 19$, which is not true.

7. Answers will vary. Sample responses:

- The output of one arrow is the input for the next arrow.
- Each arrow stands for a step in the computational process so it is reasonably easy to locate a calculation error.
- Arrow language makes writing the computation easy.
- It is easy to see subtotals with arrow language.
- It is better than using the equal sign incorrectly.

Materials transparency of the chart on Student Book page 2, optional (one per class); overhead projector, optional (one per class)

Overview Students use a chart to compute the changes in the number of passengers on the bus after various stops. They also use arrow language to describe the calculations in the chart.

About the Mathematics Arrow language is a tool that helps students to organize the order of addition and subtraction calculations. It can also be used to organize the order of multiplication and division calculations. However, arrow language is read from left to right and does not take into account that multiplication and division should be calculated before addition and subtraction. The order of operations concept is addressed in more detail in Section F of this unit.

Planning You may want to prepare a transparency of the chart on Student Book page 2 to facilitate a discussion about problems **4** and **5**. Also, briefly discuss arrow language and problem **7.**

Comments about the Problems

4. You might want to point out to students that when the number of people getting on the bus is the same as the number getting off the bus, there is no change in the number of people on the bus.

6. As discussed in problem **3**, this expression is not correct. The mathematical convention is that the quantity to the left of an equal sign is the same as the quantity to the right of the equal sign.

7. Arrow language is read from left to right. An intermediate result is the input for the next calculation. The number on the right of an arrow is not the same as the number on the left of the arrow.

Extension You might want to have students create and display their own arrow strings. Display some incorrect strings as well as correct ones. Talk about what is correct and incorrect about the strings. Also display those that use both addition and subtraction so students can see a variety of problems.

Ms. Moss had $1,235 in her bank account. Then she took out $357. Two days later, she withdrew $275 from her account.

8. Use arrow language to show what happened and how much money she had in her account at the end of the story.

John had $37 before he earned $10 for delivering newspapers one Monday. The same day, he spent $2 for an ice-cream cone. Tuesday, he visited his grandmother and earned $5 for washing her car. Wednesday, he earned $5 for baby-sitting. On Friday, he spent $2.75 for a hamburger and fries and $3.00 for a magazine.

9. a. Use arrow language to show how much money John had left.

b. Suppose John wanted to buy a radio that cost $53. Did John have enough money to buy it at any time during the week described above? If so, when?

In a region known for having lots of snow in the winter, there were 42 inches of snow on the ground one Sunday. This was what happened during the following week:

Monday	It snowed 20.25 inches.
Tuesday	It warmed up, and 8.5 inches of snow melted.
Wednesday	Two inches of snow melted.
Thursday	It snowed 14.5 inches.
Friday	It snowed 11.5 inches in the morning and then stopped.

10. How deep was the snow on Friday afternoon?

Solutions and Samples
of student work

8. $1,235 $\xrightarrow{-\$357}$ $878 $\xrightarrow{-\$275}$ $603

Ms. Moss had $603 in her account.

9. **a.** Monday: $37 $\xrightarrow{+\$10}$ $47 $\xrightarrow{-\$2}$ $45

Tuesday: $45 $\xrightarrow{+\$5}$ $50

Wednesday: $50 $\xrightarrow{+\$5}$ $55

Friday: $55 $\xrightarrow{-\$2.75}$ $52.25 $\xrightarrow{-\$3}$ $49.25

John had $49.25 left.

b. He had enough money on Wednesday, right after he earned $5 for baby-sitting.

10. The snow was 77.75 inches deep on Friday afternoon. Strategies will vary. Sample strategies:

Strategy 1

Using arrow language:

Monday: $42 \xrightarrow{+\,20.25} 62.25$

Tuesday: $62.25 \xrightarrow{-\,8.5} 53.75$

Wednesday: $53.75 \xrightarrow{-\,2} 51.75$

Thursday: $51.75 \xrightarrow{+\,14.5} 66.25$

Friday: $66.25 \xrightarrow{+\,11.5} 77.75$

Strategy 2

Some students may list the snow activity as a string of numerical operations.

$\underline{42 + 20.25} - 8.5 - 2 + 14.5 + 11.5$

$\underline{62.25 - 8.5} - 2 + 14.5 + 11.5$

$\underline{53.75 - 2} + 14.5 + 11.5$

$\underline{51.75 + 14.5} + 11.5$

$\underline{66.25 + 11.5}$

77.75

Hints and Comments

Materials newspaper and magazine articles, optional (two or three per class)

Overview Students use arrow language to solve problems about money and snowfall.

About the Mathematics The calculations described with arrow language always go in the direction of the arrows. Here the arrow strings read from left to right. Be sure students use arrow language notation correctly from the start. This will help to avoid unnecessary complications later in the unit when arrow strings are reversed and calculations go from right to left.

The calculations involve adding and subtracting decimal numbers. In the context of money and snow height, students may reinvent the rules for adding decimals. You may then have them reflect on and share the way they do these calculations.

Comments about the Problems

9. **Informal Assessment** This problem assesses students' ability to correctly use arrow strings to describe and perform a series of calculations. Watch for students who use the equal sign incorrectly, as shown in the example below:

$37 + 10 = 47 - 2 = 45 + 5 =$
$50 + 5 = 55 - 2.75 = 52.25 - 3 = 49.25$

Discuss with these students why this is an inappropriate use of the equal sign.

Writing Opportunity You may want to collect examples from newspapers or magazines in which arrow language can be used to describe the situation. Then have students write a paragraph in their journal describing how they would use arrow language to organize the information from the newspaper or magazine article.

Airline Reservations

There are 375 seats on a flight to Atlanta, Georgia that will leave on March 16th. On March 11th, 233 of the seats have been reserved. The airline continues to take reservations and cancellations until the plane takes off. If there are more seats reserved than there are seats on the plane, the airline creates a waiting list.

The table below shows what happens over the next five days.

11. Copy and complete the table.

Date	Seats Requested	Cancellations	Total Seats Reserved
3/11			233
3/12	47	0	
3/13	51	1	
3/14	53	0	
3/15	5	12	
3/16	16	2	

12. Write one or more arrow strings to show the calculations you made to complete the table.

13. When does the airline need to form a waiting list?

14. Toni, a reservations agent, suggests that, instead of arrow strings, it would be easier to add all the new reservations and then subtract all of the cancellations to get the total number of reserved seats. What are the advantages and disadvantages of her suggestion?

11.

Date	Seats Requested	Cancellations	Total Seats Reserved
3/11			233
3/12	47	0	**280**
3/13	51	1	**330**
3/14	53	0	**383**
3/15	5	12	**376**
3/16	16	2	**390**

12. Arrow strings will vary. Sample response:

3/12 $233 \xrightarrow{+47} 280 \xrightarrow{-0} 280$

3/13 $280 \xrightarrow{+51} 331 \xrightarrow{-1} 330$

3/14 $330 \xrightarrow{+53} 383 \xrightarrow{-0} 383$

3/15 $383 \xrightarrow{+5} 388 \xrightarrow{-12} 376$

3/16 $376 \xrightarrow{+16} 392 \xrightarrow{-2} 390$

13. The airline needs to begin a waiting list on March 14.

14. Answers will vary. Sample response:

One advantage is that it quickly tells you how many people are booked for the flight on the 16th. One disadvantage is that you do not know on what day they started a waiting list.

Materials transparency of the table on page 4 of the Student Book, optional (one per class); overhead projector, optional (one per class)

Overview Students complete a table, describe the calculations with arrow language, and explore a shortcut to solving the problem.

About the Mathematics The inverse operations of addition and subtraction are investigated. This lays the foundation for adding integers. Problem **14** is meant to have students investigate other ways of looking at the problem. Instead of computing one addition and one subtraction for each day, students explore totaling all the seats requested for the week and adding that total to the starting total. Then they total all the cancellations for the week and subtract that total from the sum. If information about an intermediary total is not needed, this strategy uses fewer computations and may be more efficient.

Planning You may want to prepare a transparency of the table in problem **11** to facilitate your discussion. Make sure all students understand how to use and read arrow language.

Comments about the Problems

14. This strategy uses fewer computations. However, by using this method, information about what happened each day gets lost. It depends on the question asked if this loss of information is a disadvantage.

Writing Opportunity As an extension to problem **14,** you may have students write a letter to a travel agency inquiring how it sells airline tickets and how often it oversells tickets for flights. Their letter should include a description of their solutions to the problems on this page.

Wandering Island

Wandering Island constantly changes shape. One side washes away, while on the other side, sand washes onto shore. The islanders wonder whether their island is getting larger or smaller. In 1988, the area of the island was 210 square kilometers. Since then, the islanders have kept track of the area that has washed away and the area that is added to the island.

Year	Area Washed Away (in km²)	Area Added (in km²)
1989	5.5	6.0
1990	6.0	3.5
1991	4.0	5.0
1992	6.5	7.5
1993	7.0	6.0

15. What was the area of the island at the end of 1991?

16. a. At the end of 1993, was the island larger or smaller than it was in 1988?

b. Explain or show how you got your answer to part **a.**

15. The area of the island is 209 square kilometers.

16. a. At the end of 1993, the island is smaller than it was in 1988. In 1988 it was 210 km² and in 1993 it was 209 km².

b. Explanations will vary. Some students may describe adding a column to the table to show the total change in area.

$$+ \, 0.5$$
$$- \, 2.5$$
$$+ \, 1.0$$
$$+ \, 1.0$$
$$\underline{- \, 1.0}$$
$$\text{total} \quad - \, 1.0$$

Other students may have calculated the area at the end of each year using an arrow string.

$$210 \xrightarrow{\;+\,0.5\;} 210.5 \xrightarrow{\;-\,2.5\;} 208$$
$$\xrightarrow{\;+\,1.0\;} 209 \xrightarrow{\;+\,1.0\;} 210 \xrightarrow{\;-\,1.0\;} 209$$

Overview Students calculate the change that occurs over several years in the area of an island.

About the Mathematics The problems on this page are similar to the problems on the previous pages. Information is presented in a chart and students describe the situation by using arrow language or by using the shortcut method described in problem **14.**

Comments about the Problems

15–16. These problems may be used to reinforce the use of arrow language. Students may use arrow strings to calculate the size and then compare the new size to the original size. Arrow language may be used in a variety of ways:

- one arrow for each change
- one arrow for each year
- one arrow for the total area washed away and another arrow for the total area added

Did You Know? A hurricane or major storm can cause an entire shoreline to shift. On the coasts of Georgia or Florida, it is not unusual to find the water's edge close to houses that were far from the beach before the storm. You may have students investigate how homeowners and the government work together to remedy this situation.

Fish

At the beginning of the year, Lake Mason held about 30,000 fish. During the year, 12,000 new fish hatched, and 5,000 others died naturally. Another 6,000 were caught by fishermen.

17. a. Write an arrow string to show the changes in the number of fish.

    b. What is the total change in the number of fish?

18. a. Describe the total change using only one arrow.

$$30,000 \xrightarrow{\quad ? \quad} \underline{\quad\quad}$$

    b. How did you decide what number to place over the arrow?

This string was written for another lake:

$$40,000 \xrightarrow{+\,10,000} \underline{\quad} \xrightarrow{+\,15,000} \underline{\quad} \xrightarrow{-\,8,000} \underline{\quad} \xrightarrow{\quad\quad} \underline{\quad}$$

19. What change should go above the last arrow so that the final number of fish is the same as the beginning number? Explain how you decided.

Summary

Arrow language can be helpful.

A calculation can be described with an arrow:

$$\textbf{starting number} \xrightarrow{\textbf{action}} \textbf{resulting number}$$

A series of calculations can be described by a string of arrows:

$$10 \xrightarrow{+\,6} 16 \xrightarrow{+\,3} 19 \xrightarrow{+\,3} 22 \xrightarrow{-\,4} 18$$

Summary Questions

20. a. Find the result of the following arrow string:

$$12.30 \xrightarrow{+\,1.40} \underline{\quad} \xrightarrow{-\,0.62} \underline{\quad} \xrightarrow{+\,5.83} \underline{\quad} \xrightarrow{-\,1.40} \underline{\quad}$$

    b. Make up a story that could go with the arrow string in part **a.**

21. Make up a problem that you could solve using arrow language. Then solve the problem.

22. Why is arrow language useful?

17 a.

$$30{,}000 \xrightarrow{\;+\,12{,}000\;} 42{,}000 \xrightarrow{\;-\,5{,}000\;} 37{,}000 \xrightarrow{\;-\,6{,}000\;} 31{,}000$$

b. The total change is an increase of 1,000 fish.

18. a. $30{,}000 \xrightarrow{\;+\,1{,}000\;} 31{,}000$

b. Answers will vary. Some students may add all of the increases and add all of the decreases, then find the difference between the total increase and the total decrease and adjust the total accordingly.

Others may compare 30,000 to 31,000 and realize a total increase of 1,000.

19. −17,000 should go above the last arrow. Explanations will vary. Some students may find each change and then have 57,000 after the third arrow. Next they would look at the difference between 57,000 and 40,000 to realize a change of −17,000 is needed. Other students might compare the increase to the decrease (+25,000 versus −8,000) and realize that to balance the change, a decrease of 17,000 is necessary.

20. a.

$$12.30 \xrightarrow{\;+\,1.40\;} 13.70 \xrightarrow{\;-\,0.62\;} 13.08 \xrightarrow{\;+\,5.83\;} 18.91 \xrightarrow{\;-\,1.40\;} 17.51$$

b. Stories will vary. Sample student response:

Vic had $12.30 in his pocket. His mom gave him $1.40 for bus fare. On the way to the bus stop he bought a pen for $0.62. Then he sold his lunch to Joy for $5.83. He paid the bus driver $1.40. How much did Vic have left?

21. Answers will vary. Sample response:

Fourteen people got on the empty bus at the first stop. At the second stop, two got off and eight got on. How many were still on the bus? [20 people or 21 people if you count the driver]

$$14 \xrightarrow{\;-\,2\;} 12 \xrightarrow{\;+\,8\;} 20$$

22. Answers will vary. Sample response:

Arrow language shows all the steps in order so you can find answers that are in the middle of a series of calculations.

Overview Students work with arrow language in a new context. They then read the Summary which reviews how to use arrow language.

About the Mathematics The problems on this page show how to condense an arrow string to one arrow. For example:

$$52 \xrightarrow{\;-\,8\;} 44 \xrightarrow{\;+\,3\;} 47 \xrightarrow{\;+\,8\;} 55$$

can be condensed to

$$52 \xrightarrow{\;+\,3\;} 55$$

Planning After students complete Section A, you may assign appropriate activities in the Try This! section, located on pages 54–58 of the Student Book, for homework.

Comments about the Problems

19. Encourage students to share their strategies with the class.

20. b. Many of the stories might have to do with money to make effective use of the decimals. Watch for students who use inappropriate contexts, such as people, in their problems. You may have students enter their stories in their journals.

21. The numbers that students choose to use may indicate numbers they are comfortable working with.

Extension The numbers used for the population of fish in Lake Mason were rounded. You may have students list as many reasons they can think of as to why rounded numbers were used in this context. [Answers will vary. Sample responses: It is impossible to count the exact number of fish in a lake, so an estimate is used. Exact numbers are not needed by wildlife experts as long as the fish are not on the endangered species list.] Then have students investigate the capture-tag-recapture method of estimating populations of fish and other wildlife creatures.

Source: Exploring Surveys and Information from Samples by James Landwehr, Jim Swift, and Ann Watkins (Palo Alto, California: Dale Seymour Publications, 1987).

Section Focus

Students investigate strategies for determining the correct change from a purchase and for rewriting numerical expressions to simplify calculations. The instructional focus of Section B is to:

* Determine the correct change from a purchase.
* Shorten and extend arrow strings to simplify calculations.

Planning Instruction

Day 3. Making Change		Student pages 7–11
INTRODUCTION	Problems 1 and 2	■ Discuss strategies for determining the correct change when purchasing items.
CLASSWORK	Problems 3–8	■ Use the "small-coins-and-bills-first method" to calculate change from a purchase.
HOMEWORK	Problems 9 and 10	■ Shorten addition and subtraction arrow strings.

Additional Resources: Bringing Math Home, page 21; Writing Opportunity, pages 23, 25, 27

Day 4. Skillful Computations (Continued)		Student pages 12–14
INTRODUCTION	Review homework	■ Review homework activity from Day 3.
CLASSWORK	Problems 11–15	■ Shorten and extend arrow strings to simplify calculations.
ASSESSMENT	Problems 16–19	■ Summary Questions.

Additional Resources: Assessment Opportunity, page 31; Extension, page 33; Try This! Section B, Student page 54; *Number Tools, Volume 2*, page 6

Materials

Student Resources	**Teacher Resources**	**Student Materials**
No resources required.	No resources required.	No materials required.

*See Hints and Comments for optional materials.

Concept Development

Shortening and Extending Arrow Strings

Section B introduces strategies for making change from a purchase.

On **Day 3,** students discuss strategies for determining the correct change when purchasing items. After discussing the "small-coins-and-bills-first method" to calculate the correct change from a purchase, students apply the strategy using arrow language. For homework, student shorten addition and subtraction arrow strings.

On **Day 4,** students review homework activity from Day 3. They shorten and extend arrow strings to simplify a variety of calculations. Next, they demonstrate their ability to rewrite numerical expressions to facilitate accurate and efficient calculations.

Planning Assessment

Problems 7 and 18
- Describe and perform a series of calculations using an arrow string.

Problems 16, 17, Assessment Opportunity on page 31
- Rewrite numerical expressions to facilitate calculation.

B. SMART CALCULATIONS

Making Change

Today, making change in a store is easy. You just push some buttons, and the cash register shows the amount of change due. Before computerized cash registers, however, making change was not quite so easy. Because of this, people developed a strategy for making change, which is still useful in places where computerized registers are not available.

1. a. When you buy something, how do you know if you were given the correct change?

    b. If you had to make change without a cash register, how would you do it?

Think about the following:

A total purchase is $3.70. The customer gives the clerk a $20 bill.

2. Explain how to figure the correct change without using pencil and paper or a calculator.

Solutions and Samples
of student work

1. a. Answers will vary. Some students may say that they subtract the amount from the total and then count their change to make sure they have the correct amount. Other students may estimate the correct change. Still others may suggest the *counting on* strategy, where students count the change from the total price to the amount they gave the cashier.

b. Answers will vary, but students may suggest using any of the strategies mentioned in part **a.**

2. Strategies will vary. Sample strategies:

Strategy 1

Some students may subtract mentally in steps:
$20.00 − $3.00 = $17.00.
$17.00 − $0.70 = $16.30

Strategy 2

Some students may count on from $3.70 to $20.00. The change should be a nickel to reach $3.75, a quarter to reach $4.00, a dollar to reach $5.00, a five-dollar bill to reach $10.00, and a ten-dollar bill to reach $20.00. So the total change should be a nickel, a quarter, a dollar, a five-dollar bill, and a ten-dollar bill or $0.05 + $0.25 + $1.00 + $5.00 + $10.00 = $16.30.

Overview Students investigate ways to give change when purchasing items at a checkout counter in a store.

Planning You may want to do problem **1** as a class activity. Students may then do problem **2** in pairs or in small groups. Discuss different ways to figure correct change with the class. Students may also role play both situations, one in which they use a calculator and one in which they use the "counting on" strategy.

Comments about the Problems

1. Not all students will be familiar with each of the methods mentioned by other students. If a student suggests a new strategy, have him or her show the method to the class.

2. Students might consider which of the strategies discussed in problem **1** will work best with items that do not cost much or with high-priced items. Ask students, *Which of the strategies we discussed seems to be the most efficient and easiest to use?* This question will help them to compare the various strategies and may lead them to consider the "counting on" strategy.

Bringing Math Home Have students ask their parents, brothers, or sisters what methods they use to check the change that a cashier gives them, and then have students report their findings to the class. You may also ask students if they have noticed how cashiers return change. Some cashiers place bills given them by the customers on the shelf or drawer of the register while they make change. Ask students why cashiers do this.

It would be useful to have a strategy that works in any situation. Rachel suggests beginning by estimating. "In the example," she explains, "it is easy to tell that the change will be more than $15." She says that the first step is to give the customer $15.

Rachel explains that once the $15 is given, it is as if the customer has paid only $5. "Now, only the difference between $5.00 and $3.70 must be found, and that is $1.30, or one dollar, one quarter, and one nickel."

3. Do you think that Rachel has come up with a good strategy? Why or why not?

Many people use a slightly different strategy that pays small coins and bills first. Remember, the total cost is $3.70, and the customer is paying with a $20 bill.

The clerk first gives a nickel and says, "$3.75."

Next, the clerk gives a quarter and says, "That's $4."

Then, the clerk gives a dollar and says, "That's $5."

The clerk then gives a $5 bill and says, "That makes $10."

Finally, the clerk gives a $10 bill and says, "That makes $20."

Someone thinks that this method should be called "making change to five dollars."

3. Answers will vary. Sample responses:

Rachel's method is good because she simplified the problem to work with smaller numbers. For example, by giving back $15 first, Rachel turned a $20 change-making problem into a $5 change-making problem.

Rachel's method is awkward because it requires too many steps.

Overview Students first analyze an estimation strategy for giving change and then read about a "counting on" strategy for making change.

About the Mathematics The numbers on this page have prices that end only in 5 or 0 cents, so the smallest coin used in these situations is a nickel.

Comments about the Problems

3. You might discuss students' various reasons for their answers to this question.

Did You Know? Students are probably familiar with most denominations of U.S. currency. Less common denominations include a half-dollar coin, a one-dollar coin, and a two-dollar bill. Other countries use different denominations for their currencies. For example, in The Netherlands, there is a coin with a value of 2.5 guilders and a bill with a value of 25 guilders.

Writing Opportunity Ask students to investigate money denominations from different countries. The report may be used to satisfy requirements from social studies and geography classes.

One person suggests that this method uses the fewest coins and bills. Someone else says that this could be called the "small-coins-and-bills-first method."

4. a. Does this method, in fact, use the fewest possible coins and bills?

   b. Why do you think it might also be called the small-coins-and-bills-first method?

Another customer has a total bill of $7.17 and pays with a $10 bill.

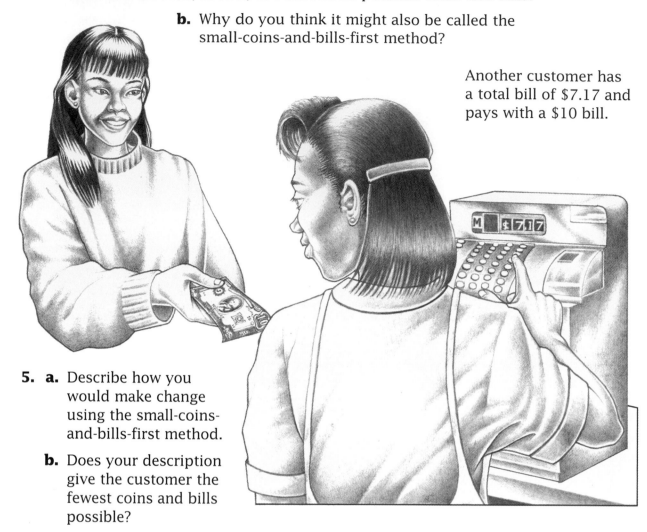

5. a. Describe how you would make change using the small-coins-and-bills-first method.

   b. Does your description give the customer the fewest coins and bills possible?

One problem with the small-coins-and-bills-first method is that you do not know the total change. Arrow language can be used to show this method of making change. Here is how the clerk could show the change for the $3.70 purchase.

$$\$3.70 \xrightarrow{+\ \$0.05} \$3.75 \xrightarrow{+\ \$0.25} \$4.00 \xrightarrow{+\ \$1.00} \$5.00 \xrightarrow{+\ \$5.00} \$10.00 \xrightarrow{+\ \$10.00} \$20.00$$

6. a. Write what the cashier would say for the third arrow above.

   b. What is the total amount of change?

   c. Write a new arrow string with the same beginning and end but with only one arrow. Explain your reasoning.

4. a. Yes, it uses 2 coins and 3 bills.

   b. You start with small change first: pennies, nickels, dimes, quarters. Then you work your way up through $1 bills, $5 bills, $10 bills, and so on.

5. a. Give three pennies to make $7.20, then one nickel to make $7.25, three quarters (or a half dollar and one quarter) to make $8.00. Then give two one-dollar bills to make $10.00.

   b. Yes, since no combination of these coins or bills can be exchanged for fewer coins or bills.

6. a. and one dollar makes five dollars.

   b. $16.30

   c. $3.70 $\xrightarrow{+\ 16.30}$ $20.00

   Explanations will vary. Some students may indicate that they subtracted $3.70 from $20 to figure out that the change is $16.30. Others might use their answer from problem **6b**.

Overview Students use the "small-coins-and-bills-first method" to solve problems. They also use arrow language to record counting change.

About the Mathematics The "small-coins-and-bills-first method" is another name for the counting on method. Using this method, a person starts with the total price and counts up to the amount given to the cashier, starting with the least-value coin or dollar. The advantage of this strategy is that the total amount of the change does not need to be computed before the cashier gives back the change.

Comments about the Problems

4. a. Students might consider why the given statement is true. You might ask, *Is there any way to use fewer nickels? dimes? Why would you want to make change using the fewest number of coins?* [Coins are heavy and most people don't want to carry large amounts of change in their pockets.]

6. After students have worked this problem, you might have them read the arrow string aloud, saying the words as a cashier might say them.

Writing Opportunity You may have students create a word problem that can be solved by using the "small-coins-and-bills-first method." Then have students show the solution to their problem. Have students write their word problem in their journals.

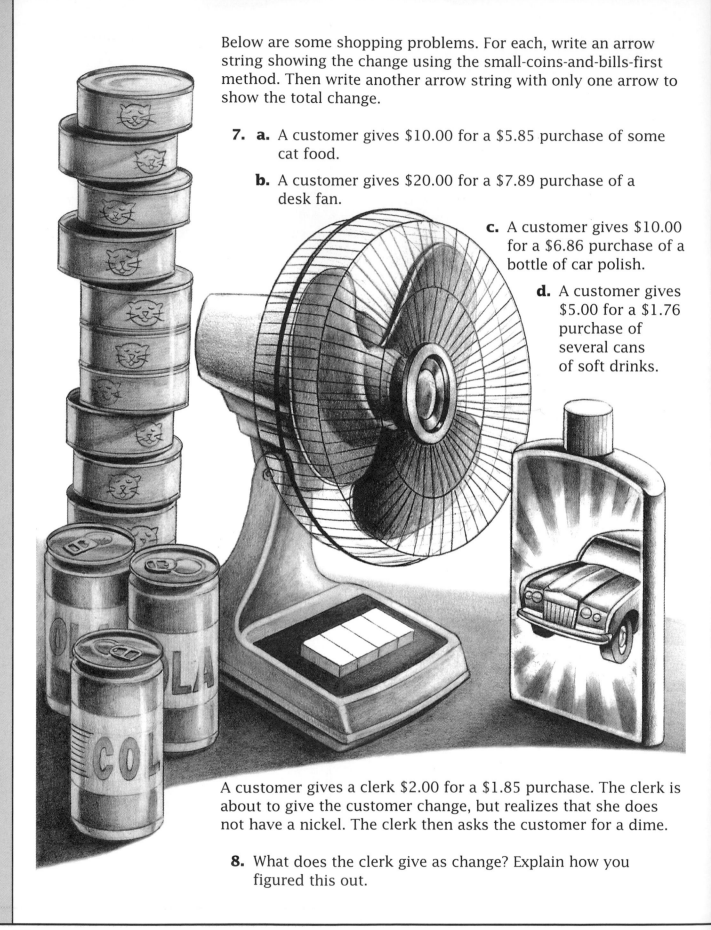

Below are some shopping problems. For each, write an arrow string showing the change using the small-coins-and-bills-first method. Then write another arrow string with only one arrow to show the total change.

7. a. A customer gives $10.00 for a $5.85 purchase of some cat food.

b. A customer gives $20.00 for a $7.89 purchase of a desk fan.

c. A customer gives $10.00 for a $6.86 purchase of a bottle of car polish.

d. A customer gives $5.00 for a $1.76 purchase of several cans of soft drinks.

A customer gives a clerk $2.00 for a $1.85 purchase. The clerk is about to give the customer change, but realizes that she does not have a nickel. The clerk then asks the customer for a dime.

8. What does the clerk give as change? Explain how you figured this out.

Solutions and Samples
of student work

7. a. $5.85 \xrightarrow{+\,0.05} \$5.90 \xrightarrow{+\,0.10} \$6.00 \xrightarrow{+\,\$4.00} \$10.00$

$\$5.85 \xrightarrow{+\,\$4.15} \$10.00$

b. $\$7.89 \xrightarrow{+\,0.01} \$7.90 \xrightarrow{+\,0.10} \$8.00 \xrightarrow{+\,\$2.00}$

$\$10.00 \xrightarrow{+\,\$10} \$20.00$

$\$7.89 \xrightarrow{+\,\$12.11} \$20.00$

c. $\$6.86 \xrightarrow{+\,0.04} \$6.90 \xrightarrow{+\,0.10} \$7.00 \xrightarrow{+\,\$3.00} \$10.00$

$\$6.86 \xrightarrow{+\,\$3.14} \$10.00$

d. $\$1.76 \xrightarrow{+\,0.04} \$1.80 \xrightarrow{+\,0.20} \$2.00 \xrightarrow{+\,\$3.00} \$5.00$

$\$1.76 \xrightarrow{+\,\$3.24} \$5.00$

8. The clerk gives back one quarter. Strategies will vary. Sample strategies:

Strategy 1

Some students may use arrow language.

$\$1.85 \xrightarrow{-\,0.10} \$1.75 \xrightarrow{+\,0.25} \$2.00$

or

$\$1.85 \xrightarrow{+\,0.25} \$2.10$

Strategy 2

Some students may reason that the change due the customer, 15¢, is only one dime short of a quarter. Having the customer pay another dime increases her debt to the customer to 25¢.

Hints and Comments

Materials play money kits, optional (one per group)

Overview Students write arrow strings to determine the correct change that should be given in different shopping problems.

Comments about the Problems

7. Informal Assessment This problem assesses students' ability to describe and perform a series of calculations by using an arrow string.

Observe whether or not students are actually using arrow language and the counting on strategy. If students are having difficulty, you might let them practice making change with play money.

8. After you discuss this situation in class, consider challenging students to play the role of a customer. Then ask them what additional money they can give the cashier to minimize the number of smaller coins they receive as change. There are many possible solutions.

Writing Opportunity Have students write in their journals about two situations at home in which they or someone in their family bought something requiring change. Ask them to describe how the change was made and how they could determine whether or not it was the correct amount.

Skillful Computations

In problem **7,** you wrote two arrow strings for the same problem. One had many arrows, and the other had only one arrow.

9. Shorten the following arrow strings so that each has only one arrow.

   a. $375 \xrightarrow{+50} \underline{\quad?\quad} \xrightarrow{+50} \underline{\quad?\quad}$ is the same as $375 \xrightarrow{?} \underline{\quad?\quad}$

   b. $158 \xrightarrow{-1} \underline{\quad?\quad} \xrightarrow{+100} \underline{\quad?\quad}$ is the same as $158 \xrightarrow{?} \underline{\quad?\quad}$

   c. $1{,}274 \xrightarrow{-1{,}000} \underline{\quad?\quad} \xrightarrow{+2} \underline{\quad?\quad}$ is the same as $1{,}274 \xrightarrow{?} \underline{\quad?\quad}$

Some of these arrow strings, such as **9b** above, are easier to use for calculation when they are long. Others, such as **9a** above, are easier to use when they are short. Sometimes an arrow string can be made easier to use by making it shorter or longer.

10. For each of the arrow strings below, make a longer string that is easier. Then use the new arrow string to find the result.

   a. $527 \xrightarrow{+99} \underline{\quad?\quad}$

   b. $274 \xrightarrow{-98} \underline{\quad?\quad}$

9. a. $375 \xrightarrow{+\,100} \underline{475}$

b. $158 \xrightarrow{+\,99} \underline{257}$

c. $1{,}274 \xrightarrow{-\,998} \underline{276}$

10. a. $527 \xrightarrow{+\,100} 627 \xrightarrow{-\,1} 626$

b. $274 \xrightarrow{-\,100} 174 \xrightarrow{+\,2} 176$

Overview Students shorten arrow strings and learn about skillful ways to break a calculation into parts to make it easier to compute.

About the Mathematics The problems on this page illustrate that calculations with numbers that are close to "friendly" numbers (multiples of 10) can be simplified with one extra step. This strategy is particularly useful when solving such problems mentally.

Comments about the Problems

9. For each problem, you may discuss which arrow string is easier to use and why.

10. Some students will rewrite 99 in problem **10a** as 100 − 1; others may rewrite 99 as −1 + 100. You may want to point out that both of these produce the same result. This is an intuitive approach to the commutative property: the order in which two numbers are added does not change the result.

11. Change each of the following calculations into an arrow string with one arrow. Then make a longer arrow string that is easier to use.

    a. 1,003 − 999

    b. 423 + 104

    c. 1,793 − 1,010

12. **a.** Guess the result of the arrow string below and then copy and complete it in your notebook.

$$273 \xrightarrow{-100} \underline{} \xrightarrow{+99} \underline{}$$

    b. If the 273 in part **a** is replaced by 500, what is the new result?

    c. What if 1,453 is used instead of 273?

    d. What if 76 is used?

    e. What if 112 is used?

    f. Use one arrow to show what happens no matter what the first number is.

           Britannica Mathematics System

11. a. $1{,}003 \xrightarrow{\;-\,999\;} 4$

$1{,}003 \xrightarrow{\;-\,1000\;} 3 \xrightarrow{\;+\,1\;} 4$

b. $423 \xrightarrow{\;+\,104\;} 527$

$423 \xrightarrow{\;+\,100\;} 523 \xrightarrow{\;+\,4\;} 527$

c. $1{,}793 \xrightarrow{\;-\,1010\;} 783$

$1{,}793 \xrightarrow{\;-\,1000\;} 793 \xrightarrow{\;-\,10\;} 783$

12. a. Estimates will vary. The correct answer is 272.

$273 \xrightarrow{\;-\,100\;} \underline{173} \xrightarrow{\;+\,99\;} \underline{272}$

b. 499

c. 1,452

d. 75

e. 111

f. $\underline{\;\;?\;\;} \xrightarrow{\;-\,1\;} \underline{\;\;?\;\;}$

Overview Students continue to investigate adding steps to calculations to make them easier to compute.

Comments about the Problems

11. These problems focus on different calculations that produce the same result and should help students begin to see how such problems could be solved mentally.

12. a. Some students may realize that subtracting 100 and adding 99 to a number produces the same result as subtracting one. Others may have to explore the problem and think about what is happening before they are ready to make that generalization.

Assessment Opportunity The following problems can be used to informally assess students' ability to rewrite numerical expressions to facilitate calculation.

1. $126 \xrightarrow{\;-\,19\;} \underline{\;\;?\;\;}$

$[126 \xrightarrow{\;-\,20\;} 106 \xrightarrow{\;+\,1\;} 107]$

2. $522 \xrightarrow{\;+\,107\;} \underline{\;\;?\;\;}$

$[522 \xrightarrow{\;+\,100\;} 622 \xrightarrow{\;+\,7\;} 629]$

3. $605 \xrightarrow{\;+\,999\;} \underline{\;\;?\;\;}$

$[605 \xrightarrow{\;+\,1000\;} 1{,}605 \xrightarrow{\;-\,1\;} 1{,}604]$

If students are having difficulty expressing numbers as friendly numbers in these calculations, you might have them work on the problems in pairs.

You can break up a number in many ways, making it easier to perform calculations. To calculate 129 + 521, you can write 521 as 500 + 21 and use an arrow string.

$$129 \xrightarrow{+\,21} \underline{150} \xrightarrow{+\,500} 650$$

Sarah computed 129 + 521 as:

$$129 \xrightarrow{+\,500} \underline{} \xrightarrow{+\,20} \underline{} \xrightarrow{+\,1} \underline{}$$

13. Is Sarah's method correct?

14. How could you rewrite 267 − 28 to make it easier to compute?

15. For each of the arrow strings below, either write a new string that will make the computation easier and explain why it is easier, or explain why the string is already the easiest it can be.

  a. $423 \xrightarrow{+\,237} \underline{}$

  b. $554 \xrightarrow{-\,24} \underline{}$

  c. $29 \xrightarrow{+\,54} \underline{} \xrightarrow{-\,25} \underline{}$

  d. $998 \xrightarrow{+\,34} \underline{}$

13. a. Yes. Sarah used the original arrow string except she reversed the order of the calculations and rewrote 21 as 20 + 1.

14. Answers will vary. Sample responses:

$$267 \xrightarrow{\;-\;20\;} 247 \xrightarrow{\;-\;8\;} 239$$

or

$$267 \xrightarrow{\;-\;30\;} 237 \xrightarrow{\;+\;2\;} 239$$

15. a. Answers will vary. Sample response:

$$423 \xrightarrow{\;+\;7\;} 430 \xrightarrow{\;+\;30\;} 460 \xrightarrow{\;+\;200\;} 660$$

This string is easier because I can add the numbers in the ones place, then add the numbers in the tens place, and finally add the numbers in the hundreds place.

b. Answers will vary. Sample response:

This string is already easy because I can easily subtract 20 from 50 to get 30, so the answer is 534. I can just do it in my head.

c. Answers will vary. Sample response:

$$29 \xrightarrow{\;-\;25\;} 4 \xrightarrow{\;+\;54\;} 58$$

This string is easier because when I subtract first, it leaves an easy number to work with.

d. Answers may vary. Sample response:

$$998 \xrightarrow{\;+\;2\;} 1,000 \xrightarrow{\;+\;32\;} 1,032$$

This string is easier because when I add 2 to 998 I get lots of zeros that are easy to work with. It's easy to add numbers to 1,000.

Materials restaurant menus, optional (one per pair)

Overview Students continue to use arrow language to simplify calculations.

Planning Emphasize to students that arrow language shows the results of each step in a sequence of operations. Also, you might discuss how arrow language can be used to calculate the net change quickly by adding or subtracting the numbers above the arrows as indicated.

Comments about the Problems

13. a. Students should learn how to express numbers as friendly numbers in several ways and to work with numbers other than those that are multiples of 10.

13. b. Students may use different number combinations or use the same number combination in a different order. You might have them check each other's arrow strings to see if they produce the same result.

15. Students who are comfortable with adding and subtracting large numbers mentally may suggest that many of these arrow strings are already as easy as they can be. The solution shown in the solution column for problem **15d** shows how numbers can be rewritten to make the sum easier to find. The other examples illustrate how using friendly numbers can make the sum easy to compute.

Extension Bring in menus from a fast-food restaurant. Have students work in pairs and pretend to order lunch or dinner for themselves and a guest, find the total amount by using arrow language, imagine paying the bill with a $20 bill, and use arrow language to determine how much change they should receive.

Summary

Sometimes an arrow string can be replaced by a shorter one that is easier to use.

$$\underline{\quad} \xrightarrow{+\,64} \underline{\quad} \xrightarrow{+\,36} \underline{\quad} \text{ becomes } \underline{\quad} \xrightarrow{+\,100} \underline{\quad}$$

Sometimes an arrow string can be replaced by a longer one to make the calculation easier without changing the result.

$$\underline{\quad} \xrightarrow{-\,99} \underline{\quad} \text{ becomes } \underline{\quad} \xrightarrow{+\,1} \underline{\quad} \xrightarrow{-\,100}$$

Summary Questions

16. Write two examples in which a shorter string would be easier to use. Be sure to include both the short and long strings for each example.

17. Write two examples in which a longer string would be easier to use. Show both the short and long strings for each example.

18. Write an arrow string to show how you would make change for the following purchases:

   a. A $6.77 purchase for which the customer pays $10.00.

   b. A $12.20 purchase for which the customer pays $20.00.

19. Explain why knowing how to shorten an arrow string can be useful in making change.

16. Answers will vary. Short strings are easier when the total of the numbers over the arrows is a multiple of 10 or a number between 1 and 10.

Sample response:

(long) $232 \xrightarrow{+\,31} 263 \xrightarrow{+\,19} 282$

(short) $232 \xrightarrow{+\,50} 282$

17. Answers will vary. Longer strings are easier when the total of the numbers over the arrows is not a multiple of 10 or a number between 1 and 10.

Sample response:

(short) $232 \xrightarrow{+\,98} 330$

(long) $232 \xrightarrow{+\,100} 332 \xrightarrow{-\,2} 330$

18. a. $\$6.77 \xrightarrow{+\,0.03} \$6.80 \xrightarrow{+\,0.20} \$7.00 \xrightarrow{+\,3.00} \$10.00$

The total amount of change is $3.23.

b. $\$12.20 \xrightarrow{+\,0.05} \$12.25 \xrightarrow{+\,0.75} \$13.00 \xrightarrow{+\,2.00}$
$\$15.00 \xrightarrow{+\,5.00} \$20.00$

The total amount of change is $7.80.

19. Explanations will vary. Sample explanation:

The shortened arrow string shows the total amount of change.

Overview Students read the Summary which reviews the two strategies used in this section to make arrow string calculations easier. They write examples to show their understanding of these strategies.

Planning After students complete Section B, you may assign appropriate activities in the Try This! section, located on pages 54–58 of the Student Book, for homework.

Comments about the Problems

16–17. **Informal Assessment** These problems assess students' ability to rewrite numerical expressions to facilitate calculation.

18. **Informal Assessment** This problem assesses students' ability to describe and perform a series of calculations by using an arrow string.

Section Focus

Students use arrow strings to represent multiplication and division calculations and to write *arrow string formulas* to represent and solve problems. The instructional focus of Section C is to:

- Use arrow strings to represent multiplication and division calculations.
- Write and evaluate arrow string formulas to solve problems.

Planning Instruction

Day 5. Supermarket		Student pages 15 and 16
INTRODUCTION	Problem 1	■ Calculate the price of produce using an arrow string.
CLASSWORK	Problems 2–5	■ Use arrow strings involving multiplication to calculate the cost of produce.
HOMEWORK	Problem 6	■ Complete a chart listing the prices of various quantities of produce.

Additional Resources: *Number Tools, Volume 2,* page 8

Day 6. Taxi Fares		Student pages 17 and 18
INTRODUCTION	Problem 7	■ Calculate the cost of a taxicab ride.
CLASSWORK	Problems 8, 10, 11	■ Use arrow strings to compare different rates for taxicab rides.
HOMEWORK	Problem 9	■ Compare different rates for a taxicab ride.

Additional Resources: Bringing Math Home, page 45; Extension, page 47

Day 7. Stacking Cups		Student pages 19–21
ACTIVITY	Problems 12–21	■ Determine the height of a stack of cups using arrow string formulas.
HOMEWORK	Problems 22–25	■ Determine the height of a stack of chairs using arrow string formulas.

Additional Resources: Assessment Opportunity, page 53

Day 8. Bike Sizes		**Student pages 22–24**
INTRODUCTION	Problems 26–28	■ Write an arrow string to represent a formula to determine the dimensions of a bike.
CLASSWORK	Problems 29–31	■ Solve problems using the bicycle height formulas.
HOMEWORK	Problems 32–34	■ Begin to investigate order of operations using arrow strings.

Additional Resources: Bringing Math Home, page 55; *Number Tools, Volume 2,* page 10

Day 9. Summary		**Student pages 25 and 26**
INTRODUCTION	Review homework	■ Review homework from Day 8.
ASSESSMENT	Problems 35–39	■ Summary Questions.

Additional Resources: Try This! Section C, Student pages 55 and 56

Materials

Student Resources

No resources required.

Teacher Resources

No resources required.

Student Materials

Quantities listed are per student, unless otherwise noted.

- Centimeter ruler
- Styrofoam® cups, four

*See Hints and Comments for optional materials.

Concept Development

Use Arrow String Formulas to Solve Problems

Section C introduces the use of arrow strings to represent multiplication and division calculations.

On **Day 5,** students calculate the price of produce using multiplication arrow strings. For homework, they complete a chart listing the prices of various quantities of produce.

On **Day 6,** students calculate the cost of a taxicab ride that has an initial fee and a price per mile. Students use *arrow string formulas* that involve addition and multiplication to determine the total cost of different rides and to compare different taxicab rates.

On **Day 7,** students measure stacks of rimmed Styrofoam® cups and write arrow string formulas to determine the total height of different stacks. They compare different formulas for solving problems involving the height of a stack of cups. Students compare arrow string formulas that look different because they organize the information from the problem in different ways, but are equivalent. For homework, students determine the height of a stack of chairs using arrow string formulas.

On **Day 8,** students write an arrow string to represent a formula for the dimensions of a bike and solve problems using bicycle height formulas. For homework, students begin to investigate the order of operations by comparing arrow strings that involve the same numbers and operations but perform the operations in a different order. Note: The rules for order of operations are formally introduced in Section D.

On **Day 9,** students review the homework from Day 8 and demonstrate their ability to use an arrow string formula to solve a problem.

Content Highlight: Arrow String Formulas

In this section, students use *arrow string formulas* to solve problems involving a series of addition and multiplication calculations. When the quantities involved in these problems change, the corresponding values in the arrow string formulas change, but the calculation process does not. For example, the process for determining the height of a stack of cups remains the same for different numbers or sizes of cups; only the number in the arrow string formula change. In addition to calculating the height of stack of cups, students can reverse the arrow string formulas to find the number of cups that would reach a certain height. Since reversing arrow strings is a relatively straightforward and intuitive process, students informally learn to solve for the input of the formula in terms of the output of the formula. In the grade 8/9 unit *Graphing Equations*, students learn the formal procedures for solving equations. In other *Mathematics in Context* units, students will use *arrow language* to write recursive and direct formulas for linear and exponential relationships.

Planning Assessment

Problems 10, 23, 24, 39

• Use and interpret a simple formula.

Problems 11, 24, 38

• Use reverse operations to find the input for a given output.

Problems 16 and 17

• Rewrite numerical expressions to facilitate calculation.

Problems 23, 24, 37

• Reason from a series of calculations to an informal formula.

Problems 23 and 24

• Use word variables to describe a formula or procedure.

Problem 36 and Assessment Opportunity on page 51

• Use a formula to solve a problem.

C. FORMULAS

Supermarket

1. Tomatoes cost $1.50 a pound. Carl buys two pounds of tomatoes.

   a. Find the total price of Carl's tomatoes.

   b. Write an arrow string that shows how you found the price.

At Veggies-R-Us, you can weigh fruits and vegetables yourself and find out how much your purchase will cost. There are buttons on the scale to indicate what is being weighed.

 APPLES

 BANANAS

 CABBAGE

 CARROTS

 CELERY

 CORN

 CUCUMBERS

 GRAPES

 GREEN BEANS

 LETTUCE

 LEMONS

ONIONS

 ORANGES

 PEAS

 PEPPERS

 POTATOES

 TOMATOES

PRICE & STICKER

Tomatoes

$1.50/lb

2.00 lb

$3.00

The scale's built-in calculator computes the purchase price and prints out a small price tag. The price tag shows what is being bought, the price per pound, how many pounds are being purchased, and the total price.

The scale, like an arrow string, takes the weight as an input and gives the price as an output.

weight $\longrightarrow$ ☐ $\longrightarrow$ price

1. a. $3.00

  b. Answers will vary. Sample responses:

$1.50 $\xrightarrow{+\ \$0.50}$ $2.00 $\xrightarrow{+\ \$1.00}$ $3.00

or

$1.50 $\xrightarrow{+\ \$1.50}$ $3.00

or

0 $\xrightarrow{+\ \$1.50}$ $1.50 $\xrightarrow{+\ \$1.50}$ $3.00

or

$1.50 $\xrightarrow{\times\ 2}$ $3.00

Overview Students use an arrow string to determine the cost of produce items in a grocery store. They then read about a computerized scale that calculates the price and prints price tags for items being weighed.

About the Mathematics Students are not expected to use an algorithm to find the product of $1.50 times 2. They may use whatever method makes sense to them, such as repeated addition ($1.50 + $1.50 = $3.00) using arrow language. Formal rules for multiplication with decimals are developed in the unit *More or Less*. In *Expressions and Formulas*, the multiplication operations involve one decimal factor and one whole number factor.

The terms *input* and *output* are commonly used to describe calculations using arrow strings. Within the context on this page, the scale takes the weight of the produce as the input and prints out the price as an output. Be sure students are familiar with these terms.

Planning You might introduce the context of this section with a brief discussion about scales equipped with a calculator for computing prices.

Comments about the Problems

1. Students should begin to think about a procedure for calculating various weights (input) of produce to determine the prices (output) as a lead-in to thinking about using formulas to perform such calculations.

Did You Know? Scales like the one described on this page can be found in many grocery stores. Advanced machines read the bar code on each item that can be scanned at the checkout counter so that the cashier does not have to enter in the price manually. These scanners also keep track of how much of each fruit and vegetable is bought. This helps the store owner keep an accurate inventory of each produce item.

2. Find the price for each of the following weights of tomatoes using the following arrow string:

$$\text{weight} \xrightarrow{\times\ \$1.50} \text{price}$$

a. 4 lb $\xrightarrow{\times\ \$1.50}$ ___?___

b. 3 lb $\xrightarrow{\times\ \$1.50}$ ___?___

c. 0.5 lb $\xrightarrow{\times\ \$1.50}$ ___?___

d. 2.5 lb $\xrightarrow{\times\ \$1.50}$ ___?___

The prices of other fruits and vegetables are calculated in the same way. Green beans cost $0.90 per pound.

GREEN BEANS

3. Write an arrow string to show the calculation the scale would use for green beans.

GRAPES

The calculation for the price of grapes is given by this arrow string:

$$\text{weight} \xrightarrow{\times\ \$1.70} \text{price}$$

4. Draw price tags for the following purchases:

a. 2 lb of grapes

b. 3 lb of green beans

c. 6 lb of tomatoes

The Corner Store does not have a fancy scale. The price of tomatoes there is $1.20 per pound. Siu bought some tomatoes, and her bill was $6.

5. What was the weight of the tomatoes Siu bought? How did you calculate this?

The manager of the Corner Store wants customers to be able to estimate the prices of their purchases. She posts a chart next to a regular scale.

6. a. Help the manager by copying and completing the chart below.

Weight	Tomatoes $1.20/lb	Green Beans $0.80/lb	Grapes $1.90/lb
0.5 lb			
1.0 lb			
2.0 lb			
3.0 lb			

b. The manager wants to add more rows to the chart. Add a row to your chart to show the prices for 2.5 pounds of each item.

c. Add and complete at least three more rows to the chart.

2. a. $6.00

   b. $4.50

   c. $0.75

   d. $3.75

3. weight $\xrightarrow{\times\ \$0.90}$ price

4. a.

> ### Grapes
> $1.70/lb
> 2.00 lb
> $3.40

   b.

> ### Green Beans
> $0.90/lb
> 3.00 lb
> $2.70

   c.

> ### Tomatoes
> $1.50/lb
> 6.00 lb
> $9.00

5. Siu bought 5 pounds of tomatoes. Strategies will vary. Students may use division ($6.00 ÷ $1.20), trial and error, or repeated addition: $1.20 + $1.20 + $1.20 + $1.20 + $1.20 = $6.00; Since $1.20 was added 5 times, Siu bought 5 lb of tomatoes.

6. a–b.

	Tomatoes $1.20/lb	Green Beans $0.80/lb	Grapes $1.90/lb
0.5 lb	$0.60	$0.40	$0.95
1.0 lb	$1.20	$0.80	$1.90
2.0 lb	$2.40	$1.60	$3.80
3.0 lb	$3.60	$2.40	$5.70
2.5 lb	$3.00	$2.00	$4.75

   c. Answers will vary. Sample responses:

	Tomatoes $1.20/lb	Green Beans $0.80/lb	Grapes $1.90/lb
8.0 lb	$9.60	$6.40	$15.20
5.0 lb	$6.00	$4.00	$9.50
4.0 lb	$4.80	$3.20	$7.60

Materials calculators, optional (one per student)

Overview Students compute prices for different purchases of fruit and vegetables.

Comments about the Problems

   2. Encourage students to devise their own strategies to solve these problems. For example, part **d** could be considered the result of subtracting the answer to part **c** from the answer to part **b.**

   4. Students may use repeated addition or any other strategy that makes sense to them to find the products. For example, to find the price of 3 lb of green beans in part **b,** students may reason that the beans are 10 cents less than $1. For 3 lb, they would spend $3 minus 3 times 10 cents, or $3.00 − $0.30 = $2.70.

   5. In this problem, students are given the cost per pound and the total cost of some tomatoes and are asked to work backwards or reverse the process to find the weight of the tomatoes. Some may use a "guess and check" strategy, using values for the weight until the cost reaches $6.00. If they do this, ask them how they knew to refine their guesses to begin approaching the answer. Other students may repeatedly subtract $1.20 from $6.00 to determine the weight.

   6. Homework Some students may see this chart as a ratio table and use the patterns in the rows and columns to compute their answers.

TAXI FARES

In some cabs, the price for the ride is shown on a meter. At the Rainbow Cab Company, the price increases during the ride depending on the distance traveled. There is a starting amount you have to pay no matter how far you go, as well as a price for each mile you ride. The Rainbow Cab Company charges the following rates:

The starting amount is $2.00.
The amount per mile is $1.50.

7. What is the price for each of these rides?

    a. from the stadium to the railroad station: 4 miles

    b. from a suburb to the center of the city: 7 miles

    c. from the convention center to the airport: 20 miles

The meter has a built-in calculator to find the price of a ride. The way the meter works can be described with an arrow string.

8. Which of these strings will give the correct price? Explain your answer.

$$\text{number of miles} \xrightarrow{\times \$1.50} \underline{\quad} \xrightarrow{+ \$2.00} \text{total price}$$

$$\$2.00 \xrightarrow{+ \text{ number of miles}} \underline{\quad} \xrightarrow{\times \$1.50} \text{total price}$$

$$\text{number of miles} \xrightarrow{+ \$2.00} \underline{\quad} \xrightarrow{\times \$1.50} \text{total price}$$

7. a. $8.00. Students should use a strategy to multiply 4 × $1.50, and then add $2.00 to the product.

b. $12.50. Students should use a strategy to multiply 7 × $1.50, and then add $2.00 to the product.

c. $32.00. Students should use a strategy to multiply 20 × $1.50, and then add $2.00 to the product.

8. The first string gives the correct price. Explanations will vary. Students may realize from working problems **7a–7c** that they must multiply the amount per mile by the number of miles and then add $2.00 to the product. Other students may reason that you cannot add the number of miles to the starting amount, so the second and third strings are incorrect.

Overview Students calculate the prices for a number of cab rides and describe the calculations using arrow strings.

About the Mathematics Arrow strings are used to describe calculation rules. The units that correspond to various numbers in the arrow string, such as 1.2 miles or $1.50 per mile, need not be consistent throughout. The order of operations is informally introduced here in problem **8.** This concept is studied in greater depth later in this section and in the following sections.

Planning You may want to introduce the concept of taxicab fares to students as some may be unfamiliar to them.

Comments about the Problems

7. In this problem, students work with both multiplication and addition in the same problem for the first time. Encourage students to consider the context of this situation as a way to decide the order in which the operations are performed.

8. It is important that students understand that only one of the arrow strings is correct. Students should recognize that it does not make sense to add dollar amounts to miles. You might have students check their answers by using the distances and answers from problem **7.**

Bringing Math Home You may ask students to discuss the contexts of the scale at the grocery store and taxicab fares with their families. Students can ask their parents if they have grocery store receipts that could be brought to class. Students could then check the unit prices and total cost for various produce items on each receipt for accuracy. Students could also ask parents to recall various rides made in taxicabs and the total cost of each ride.

The Rainbow Cab Company has changed its rates. The new prices can be found using the following string:

$$\text{number of miles} \xrightarrow{\times\ \$1.30} \underline{\quad\quad} \xrightarrow{+\ \$3.00} \text{total price}$$

9. Is a cab ride now cheaper or more expensive than it was before?

10. Use the new rate to find the price for each of the following rides:

  a. from the stadium to the railroad station: 4 miles

  b. from a suburb to the center of the city: 7 miles

  c. from the convention center to the airport: 20 miles

  d. Compare your answers before the rate change (from problem **7**) with those after the rate change. Did you answer problem **9** correctly?

11. After the company changed its rates, George slept through his alarm and had to take a cab to work. He was surprised at the cost: $18.60!

  a. Use the new rate to find out how far it is from George's home to work.

  b. Write an arrow string to show your calculations for part **a.**

The arrow string for the price of a taxi ride tells how to find the price for any number of miles.

$$\text{number of miles} \xrightarrow{\times\ \$1.30} \underline{\quad\quad} \xrightarrow{+\ \$3.00} \text{total price}$$

An arrow string that tells how to do a specific type of calculation is called a *formula.*

9. Answers will vary, depending on the length of the cab ride. For rides less than 5 miles, the prices have increased; for a 5-mile ride, the prices are the same; and for rides longer than 5 miles, the prices have decreased.

10. a. $8.20

    b. $12.10

    c. $29.00

    d. The 4-mile ride became more expensive; the 7-mile and 20-mile rides became less expensive.

11. a. 12 miles. Strategies will vary. Sample strategies:

Strategy 1

$18.60 − $3.00 = $15.60; $15.60 ÷ $1.30 = 12

Strategy 2

First subtract $3.00 from $18.60, since the taxi fare rate includes a $3.00 starting amount ($18.60 − $3.00 = $15.60). The price per mile is $1.30, so you pay $13 for 10 miles and another $2.60 for 2 miles. So, George traveled a distance of 12 miles for $15.60.

    b. Answers will vary. Sample solution:

$$\$18.60 \xrightarrow{\;-\,3.00\;} \$15.60 \xrightarrow{\;\div\,1.30\;} 12$$

Overview Students continue calculating taxicab fares for various distances. They also learn that arrow strings that describe a series of calculations that can be used to solve similar problems are known as *formulas.*

About the Mathematics Formulas come in various forms, but all formulas produce an output when the input is entered. A formula usually describes a relationship between two or more variables. In the context of the taxicab fares, the two variables are mileage and the price per mile.

Planning You may want to discuss the definition of the term *formula*, since it will be used throughout the rest of the unit.

Comments about the Problems

9. Students may answer this question intuitively or they may systematically check fares for specific distances from other problems. At this point, it is not important that students recognize the correct answer.

10. Informal Assessment This problem assesses students' ability to use and interpret a simple formula. Encourage students to share strategies for calculating 20 × $1.30 for part **c.**

11. Informal Assessment This problem assesses students' ability to use reverse operations to find the input for a given output. Students should be able to explain how they computed the distance. Some may use a guess-and-check method while others may make a table and look for patterns. Still other students may create notation for reversing an arrow string.

Extension You might have students determine for what distances the new and old fares are the same. [The rates are the same for 5 miles which costs $9.50 using either fare rate.]

You might also ask students to find which fare rate is less expensive for a long ride. Students could construct a line graph that shows the two fare rates on the same axes. Discuss what the graph shows for a trip of 20 miles. [The rates are the same. The first fare rate is less expensive for trips less than 20 miles; the second fare rate is less expensive for trips over 20 miles.]

Activity

Stacking Cups

Materials:

Each group will need a centimeter ruler and at least four cups that are the same. Plastic cups from sporting events or fast-food restaurants work well.

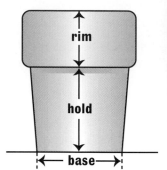

12. Measure and record the following:

- the total height of a cup
- the height of the rim
- the height of the hold

> (*Note:* The *hold* is the distance from the bottom of the cup to the bottom of the rim.)

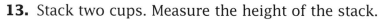

13. Stack two cups. Measure the height of the stack.

14. a. Without measuring, guess the height of a stack of four cups.

b. Write down how you made your guess. With a partner, share your guess and how you arrived at it.

c. Make a stack of four cups and measure it. Was your guess correct?

15. Use the method you discovered in problem **14** to calculate the height of a stack of 17 cups. Describe your calculation with an arrow string.

16. Choose two different numbers of cups to put in a stack. For each number, calculate the height. Then make a stack and check your calculation.

17. a. There is a space under a counter where cups will be stored. The space is 50 centimeters high. How many cups can be stacked to fit under the counter?

b. Use arrow language to explain how you found the answer.

Solutions and Samples
of student work

12. Answers will vary, depending on the cups students use. Sample measurements: The cup is 7.5 cm, the rim is 2.5 cm, and the hold is 5 cm.

13. 10 centimeters (based on sample measurements for problem **12**). The stack will measure two rims and one hold high, or 2 × 2.5 cm + 5 cm = 10 centimeters.

14. a. Answers will vary.

    b. Answers will vary. Sample response is based on measurements for problem **12:**

       I figured there would be four rims that are 2.5 cm each which is 10 cm, and one hold that is 5 cm. Since 10 cm + 5 cm = 15 cm, I guessed the height would be 15 centimeters.

    c. Answers will vary. Small discrepancies may be due to errors in measurement. Sometimes stacking increases the expected height slightly because the design of the cup prevents the rims from meeting perfectly.

15. Answers will vary. Sample solution is based on measurements for problem **12:**

$$\text{weight} \xrightarrow{\times \$0.90} \text{price}$$

16. Answers will vary. Sample responses are based on measurements for problem **12:**

Number of Cups	Height (cm)
1	7.5
2	10
3	12.5
4	15

17. a. The number of cups that would fit exactly is 18, based on the measurements for problem **12**. Some students may say 17, to facilitate moving the stack on and off the shelf. To solve this problem, students may have used guess and check, continued the table, or used repeated addition or subtraction.

    b. Answers will vary. Sample response based on measurements for problem **12:**

$$50 \text{ cm} \xrightarrow{-\ 5 \text{ cm}} 45 \text{ cm} \xrightarrow{\div\ 2.5} 18$$

Hints and Comments

Materials Styrofoam® cups (four per group); centimeter rulers (one per group)

Overview Students stack cups and determine the height of the stack using arrow language formulas.

Planning Please allow students to actually do this activity (which will take about 20 to 30 minutes) in class. Be sure the cups have a measurable rim and that they can be stacked.

Comments about the Problems

12. You may need to discuss the meanings of the terms *hold* and *rim* for some students.

13. You may need to demonstrate how to measure the height of the stack.

14. Some students may calculate an accurate answer while others may guess or estimate.

15. If students need a hint, you may suggest that they make a table to organize their work and to help them find patterns.

16. Informal Assessment This problem assesses students' ability to rewrite numerical expressions to facilitate calculation.

    If students are having difficulty, you may point out that knowing something about the measurements of the parts (the hold and rim of one cup) can help determine how high the stack will be.

17. Informal Assessment This problem assesses students' ability to rewrite numerical expressions to facilitate calculation.

    In this problem, students can work backwards or they can keep building a table of values until they reach a height of 50 centimeters. Some students may guess a certain number of cups, then check their guess until they find the correct solution.

Sometimes a formula is useful. You can write a formula to find the height of a stack of cups if you know the number of cups.

18. Complete the following arrow string for a formula using the number of cups as the input and the height of the stack as the output.

number of cups $\xrightarrow{\text{?}}$ _?_ $\xrightarrow{\text{?}}$ height of stack

Suppose a class down the hall has *different* cups. The students use the following formula for finding the height of a stack of their cups:

number of cups $\xrightarrow{-1}$ ____ $\xrightarrow{\times 3}$ ____ $\xrightarrow{+15}$ height of stack

19. a. How tall is a stack of 10 of these cups?

   b. How tall is a stack of five of these cups?

   c. Sketch one of the cups from this class.
   Label your drawing with the correct height.

   d. Explain what each of the numbers in the formula represents.

Now consider the following formula:

number of cups $\xrightarrow{\times 3}$ ____ $\xrightarrow{+12}$ height

20. Could this formula be for the same cup from Problem **19?** Explain.

21. These cups also need to be stored in a space 50 centimeters high. How many of these cups can be placed in a stack? Explain how you found your answer.

18. This arrow string is based on the measurements for problem **12.**

Number $\xrightarrow{\ \times\,2.5\ }$ Height $\xrightarrow{\ +\,5\text{ cm}\ }$ Height
of Cups $\qquad\qquad$ of Rims $\qquad\qquad$ of Stack

19. a. 42 cm

$$10 \xrightarrow{\ -\,1\ } 9 \xrightarrow{\ \times\,3\ } 27 \xrightarrow{\ +\,15\ } 42$$

b. 27 cm

$$5 \xrightarrow{\ -\,1\ } 4 \xrightarrow{\ \times\,3\ } 12 \xrightarrow{\ +\,15\ } 27$$

c.

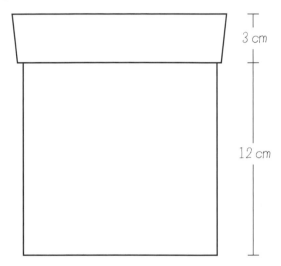

3 cm

12 cm

d. Answers will vary. Students should experiment with the numbers from parts **a** and **b** until they recognize that each cup is 15 cm high and has a hold of 12 cm and a rim of 3 cm.

20. Yes. Explanations will vary, but students may reason that the formulas are for the same cups since they both give the same results.

21. 12 cups will fit in a space that is 50 centimeters high. Explanations will vary. Sample explanation:

$$50 \text{ cm} \xrightarrow{\ -\,12\ } 38 \text{ cm} \xrightarrow{\ \div\,3\ } 12\tfrac{2}{3} \text{ cups or } 12 \text{ cups}$$

Materials Styrofoam® cups (four per group); centimeter rulers (one per group)

Overview Students compute the heights of stacks of different kinds of cups. They also determine whether or not a new formula could be used to determine the height of a stack of cups.

About the Mathematics Most formulas can be written in more than one way. The form of a formula may depend upon the purpose for which it is used. On this page, students compare two formulas for the same kind of cup. The formula in problem **19** uses the height of the cup (15 centimeters) and the rim of the cup (3 centimeters). The formula in problem **20** uses the hold (12 centimeters) and the rim of the cup (3 centimeters). Because they use different measurements, the formulas appear to be different. However, because the information used in both is from the same cup, the formulas are equivalent to each other.

Comments about the Problems

19. In part **c,** in order to sketch and label their cup, students may have difficulty trying to determine what the −1 in the arrow string represents. Direct students to first determine the meaning behind the number 15.

20. Encourage students to use examples to explain how the two formulas relate to each other. Refer to the comments in the About the Mathematics section on this page.

Stacking Chairs

The picture below shows a stack of chairs. Notice that the height of one chair is 80 centimeters, and a stack of two chairs is 87 centimeters high.

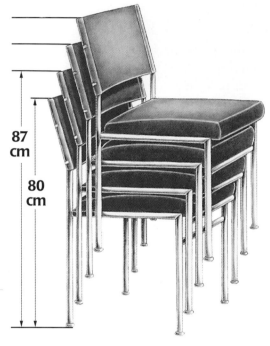

87 cm

80 cm

Damian suggests that the following formula can be used to find the height of a stack of these chairs:

number of chairs $\xrightarrow{-1}$ ____ $\xrightarrow{\times 7}$ ____ $\xrightarrow{+80}$ height

22. Explain what each of the numbers in the formula represents.

23. Alba thinks that a formula like this would do just as well:

number of chairs $\xrightarrow{\times}$ ____ $\xrightarrow{+}$ height

a. What numbers would Alba use in her formula? Explain how you determined these numbers.

b. Alba thought about making a formula like this:

number of chairs $\xrightarrow{+}$ ____ $\xrightarrow{\times}$ height

Will this work? Why or why not?

24. These chairs are used in an auditorium and sometimes have to be stored underneath the stage. The storage space is 116 centimeters high.

a. How many chairs can be put in a stack that will fit in the storage space?

b. Describe your calculation with an arrow string.

25. For another style of chair, there is a different formula:

number of chairs $\xrightarrow{\times 11}$ ____ $\xrightarrow{+54}$ height

a. How are these chairs different from the first ones?

b. If the storage space were 150 centimeters high, would the following arrow string give the number of chairs that would fit? Why or why not?

$150 \xrightarrow{\div 11}$ ____ $\xrightarrow{-54}$ ____

22. The 80 represents the height of the first chair. Seven centimeters is the height added for each additional chair. One subtracted from the number of chairs is the number of chairs that are added to the first chair in the stack.

23. a. She would use × 7 and + 73 above the arrows. Explanations will vary. Sample explanation:

I wrote "x 7" at the beginning of the formula because each chair adds 7 cm to the height of the stack. I wrote "+ 73" next to show the height of the first chair (80 cm) minus the 7 cm that was already added in the first step.

b. No, it will not produce the same result because the order of the addition and the multiplication has been switched.

24. a. 6 chairs. Strategies will vary. Some students might do calculations in reverse: subtract 73 cm from 116 cm and divide by 7.

116 cm − 73 cm = 43 and 43 ÷ 7 is a little more than 6 chairs. Round down to 6 because more than 6 chairs will be too tall for the space.

Others may make a table or simply make a guess.

b. Answers and strategies will vary. Sample strategy:

$$80 \text{ cm} \xrightarrow{+7 \text{ cm}} 87 \text{ cm} \xrightarrow{+7 \text{ cm}} 94 \text{ cm} \xrightarrow{+7 \text{ cm}}$$

$$101 \text{ cm} \xrightarrow{+7 \text{ cm}} 108 \text{ cm} \xrightarrow{+7 \text{ cm}} 115 \text{ cm}$$

25. a. These chairs are 65 cm tall and when they are stacked, each chair adds 11 cm to the stack.

b. No, the arrow string will not give the correct number of chairs. The order of the arrows must be reversed to get the correct answer.

$$150 \text{ cm} \xrightarrow{-54 \text{ cm}} 96 \text{ cm} \xrightarrow{\div 11 \text{ cm}} 8 \frac{8}{11} \text{ or 8 chairs}$$

Overview Students determine the height of a stack of chairs using an arrow string formula.

Comments about the Problems

23. Informal Assessment This problem assesses students' ability to use and interpret simple formulas, reason from a series of calculations to an informal formula, and use word variables to describe a formula or procedure.

Suggest that students try some numbers in both formulas to see how they work. You may need to remind them of the two different formulas they worked with in the Stacking Cups activity.

24. Informal Assessment This problem assesses students' ability to use and interpret simple formulas, use reverse operations to find the input for a given output, reason from a series of calculations to an informal formula, and use word variables to describe a formula or procedure. Encourage students to share their strategies with the class.

Assessment Opportunity You may want to use the following problems to assess students' ability to write and use a formula to solve a problem, and to use reverse operations to find the input for a given output.

1. One of the options offered by the phone company is a fee of $8 a month plus a charge of $0.15 a call.

a. Write a formula to determine how much your phone bill would be each month using that plan.

Number of $\xrightarrow{\times \$0.15}$ _____ $\xrightarrow{+ \$8}$ Total Phone Bill
phone calls

b. Use your formula to find how much you would pay if you made 24 calls during the month. [$11.60]

c. If Sam budgeted $20 per month for his phone bills, how many telephone calls could he make? Explain how you found your answer. [80 calls; Explanations may vary. Sample explanation: I subtracted the $8 fee from $20 and divided that amount by $0.15.]

Bike Sizes

You have discovered some formulas yourself. On the next two pages, you will use formulas that other people have developed.

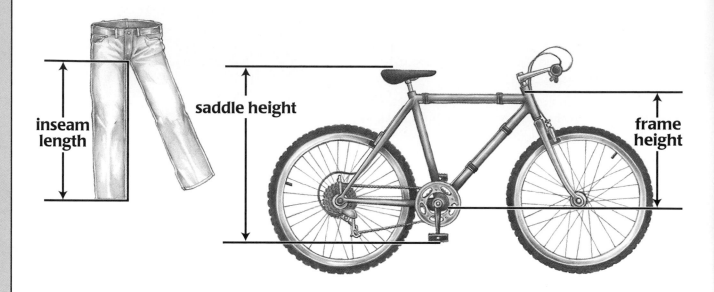

Stores that sell bicycles use formulas to find the best saddle and frame heights for each customer. One number used in these formulas is the inseam of the cyclist. This is the length of the cyclist's leg, measured in centimeters along the inside seam of the pants. The saddle height is calculated with the following formula:

> inseam (in cm) × 1.08 = saddle height (in cm)

26. Write an arrow string for this formula.

The formula for the frame height is as follows:

> inseam (in cm) × 0.66 + 2 cm = frame height (in cm)

27. Write an arrow string for this formula.

26. inseam $\xrightarrow{\times\ 1.08}$ saddle height

27. inseam $\xrightarrow{\times\ 0.66}$ _____ $\xrightarrow{+\ 2}$ frame height

Materials calculators, optional (one per student); bicycle, optional (one per class)

Overview Students investigate two formulas used for determining the correct bicycle dimensions based on a person's inseam length.

About the Mathematics Although mathematical formulas are usually written with the output to the left of the equal sign, the formulas on this page are written in the reverse order (with the output on the right), to coincide with the arrow string notation.

The calculations on this page involve multiplying whole numbers by decimal numbers having two decimal places. Some students may struggle with these calculations. Remind these students of the money context to facilitate the calculations. Also, calculations may be simplified by changing the units of measurement. For example, dollars can be converted to cents and meters converted to centimeters. Converting between different units within the same measurement system is introduced in the unit *Measure for Measure*.

Planning You may want to bring a bicycle into the classroom so that students can take some actual measurements.

Comments about the Problems

27. Calculating the saddle height for a few different inseams may be helpful before students try to write the arrow string for this formula.

Bringing Math Home You might have students bring these two formulas home and share them with their families. Suggest that students have each member of the family measure their pant inseam length to determine if each person's bicycle is the correct size. Have students share their results in class.

Formulas are often written with the result first, for example:

> saddle height (in cm) = inseam (in cm) × 1.08
>
> frame height (in cm) = inseam (in cm) × 0.66 + 2 cm

28. Are the arrow formulas you wrote (for problems **26** and **27**) correct for these rewritten formulas above?

29. Carlie has an inseam of 70 centimeters. What frame and saddle heights does she need?

30. Look at the bike pictured below.

  a. What is the frame height?

  b. What is the saddle height?

  c. Do both of these numbers correspond to the same inseam length? How did you find your answer?

31. If you were buying a bike, what frame and saddle heights would you need?

28. Yes. They produce the same result. The output appears on the left of the equal sign instead of on the right.

29. The saddle height should be 75.6 centimeters. The frame height should be 48.2 centimeters.

Strategies will vary. Sample strategy:

$$\text{inseam} \xrightarrow{\times 1.08} \text{saddle height}$$

$$70 \text{ cm} \xrightarrow{\times 1.08} 75.6 \text{ cm}$$

$$\text{inseam} \xrightarrow{\times 0.66} \underline{\quad} \xrightarrow{+ 2} \text{frame height}$$

$$70 \text{ cm} \xrightarrow{\times 0.66} 46.2 \xrightarrow{+ 2} 48.2 \text{ cm}$$

30. a. The frame height is 54 centimeters.

b. The saddle height is 81 centimeters.

c. These numbers do not correspond with the same inseam height. The given saddle height (81 cm) corresponds to a person with a 75 cm inseam, and the frame height (54 cm) corresponds to a person with a 78.8 cm inseam.

Strategies will vary. Some students may use a guess and check strategy, make a table, or use information from other problems.

31. Answers will vary, depending on students' inseam length. Students should measure their inseam and then compute the saddle height and the frame height using the given formulas.

Materials calculators, optional (one per student)

Overview Students continue solving problems using the bicycle height formulas.

About the Mathematics Students compare two formulas with their corresponding arrow strings. Help students to realize that the information on the left of the equal sign can be switched with the information on the right of the equal sign. For example, the word formula saddle height = inseam × 1.08 means the same as inseam × 1.08 = saddle height.

Comments about the Problems

30. c. Some students may use repeated subtraction to find that a saddle height of 81 centimeters corresponds with an inseam of 75 centimeters. Students can then use the inseam of 75 centimeters to find the frame height and see that it is not 54 centimeters. Other students may find the inseam for a frame height of 54 centimeters to be 75 centimeters by using the strategy described in the solution column. Then students may find that the corresponding inseam is about 79 centimeters. An inseam of 79 centimeters gives a saddle height of about 85 centimeters. So, the two numbers do not correspond to the same inseam length.

31. This problem requires students to use their measuring skills as well as the formulas developed so far. One way to measure a person's inseam is from the outside of the hip to the ankle.

Are They the Same?

32. Below are two pairs of arrow strings. They look similar, but are they really the same? Try some different numbers for the input. Then explain why you think the arrow strings in each case are the same or different.

a. input $\xrightarrow{+7}$ ____ $\xrightarrow{-2}$ output

input $\xrightarrow{-2}$ ____ $\xrightarrow{+7}$ output

b. input $\xrightarrow{\times 2}$ ____ $\xrightarrow{\div 10}$ output

input $\xrightarrow{\div 10}$ ____ $\xrightarrow{\times 2}$ output

33. Compare the two strings below and decide whether they are the same or different.

input $\xrightarrow{\times 4}$ ____ $\xrightarrow{+3}$ output

input $\xrightarrow{+3}$ ____ $\xrightarrow{\times 4}$ output

34. Write about what you discovered while working on problems **32** and **33**. Think of a rule that explains what you found.

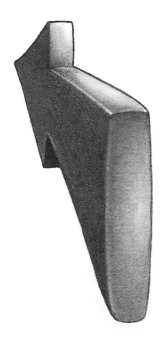
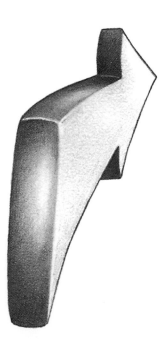

32. a. Yes. Both arrow strings produce the same result for any number used as the input.

b. Yes. Both arrow strings produce the same result for any number used as the input.

33. These arrow strings are not the same because they produce different results using the same input. The result of the second arrow string is consistently 9 units greater than the first. Students should choose various numbers to test the two arrow strings:

$$5 \xrightarrow{\times 4} 20 \xrightarrow{+3} 23$$

$$5 \xrightarrow{+3} 8 \xrightarrow{\times 4} 32$$

34. Answers will vary. Students should note that addition and subtraction can be done in any order. Multiplication and division can also be done in any order. However, when addition or subtraction is combined with multiplication or division, the order of operations is important because switching the order of operations results in different answers.

Overview Students compare different arrow strings to determine if they are equivalent expressions.

About the Mathematics The order of operations is implicitly introduced in the problems on this page. According to the order of operations, subtraction and addition can be done in any order. Multiplication and division can also be done in any order. However, when combining these operations, multiplication and division are calculated before addition and subtraction. The rules for the order of operations are formalized in Section D.

Comments about the Problems

32. Students should support their answers with examples. They may be able to generalize that since addition and subtraction (and multiplication and division) are opposite operations, the order in which they are performed does not matter.

33. This problem does not address calculations that involve all four arithmetic operations, so do not expect students to generalize about such calculations at this time.

34. Most students will not be ready to formalize the rule for the order of operations at this point in the unit.

Summary

A formula shows a procedure that can be used over and over again for different numbers in the same situation.

One formula used for bikes is:

inseam (in cm) × 0.66 + 2 cm = frame height (in cm)

or, written with the result first,

frame height (in cm) = inseam (in cm) × 0.66 + 2 cm

Many formulas can be described with arrow strings, for example:

inseam $\xrightarrow{\times\,0.66}$ ____ $\xrightarrow{+\,2}$ frame height

Sometimes it is possible to change the order of the arrows in a string. If a problem has only addition and subtraction or only multiplication and division, the order can be changed, and the result will stay the same.

Summary Questions

35. Why is it useful to write a formula as an arrow string?

35. Answers will vary. Sample responses:

I can do the calculations in steps, and that makes it easier.

You get results in the middle of the calculation.

I can recognize the calculations better in an arrow string than in a formula.

Overview Students read the Summary which reviews how formulas can be used to describe repeated calculations for problems within the same context. They then compare a word formula to an arrow string.

Summary Questions, continued

Mr. Macker is a band teacher who travels to different schools. Every day, he must decide how long it will take him to get to school and prepare to teach. He knows that he must allow 2 minutes for every mile between his home and a school. He also allows 15 minutes to set up the room before students arrive.

36. How much time should Mr. Macker allow if a school is 12 miles from his home?

37. Write a formula that Mr. Macker could use to find the time needed each day.

38. One day Mr. Macker decides that he needs 73 minutes. How far does he live from that school? Explain how you found your answer.

39. Which of the arrow strings labeled **a** and **b** will give the same result as the arrow string below? Explain.

$$\text{input} \xrightarrow{+2} \underline{\quad} \xrightarrow{+3} \underline{\quad} \xrightarrow{\times 4} \text{output}$$

a. $\text{input} \xrightarrow{+2} \underline{\quad} \xrightarrow{\times 4} \underline{\quad} \xrightarrow{+3} \text{output}$

b. $\text{input} \xrightarrow{+3} \underline{\quad} \xrightarrow{+2} \underline{\quad} \xrightarrow{\times 4} \text{output}$

36. 39 minutes. Strategies will vary. Sample strategy:

$$12 \xrightarrow{\times 2 \text{ min}} 24 \text{ min} \xrightarrow{+ 15 \text{ min}} 39 \text{ min}$$

37. number of minutes =
2 minutes/mile × miles + 15 minutes

or

2 minutes/mile × miles + 15 minutes = number of minutes

38. 29 miles. Strategies will vary. Sample strategies:

Strategy 1

Students may reverse the calculations.

73 − 15 = 58, then 58 ÷ 2 = 29

Strategy 2

Students may use a reverse arrow string

$$73 \xrightarrow{- 15} 58 \xrightarrow{\div 2} 29 \text{ minutes}$$

Strategy 3

Some students may create a table and try different input numbers.

miles	2 x miles	+ 15	
12	24	39	too low
15	30	45	too low
30	60	75	too high
29	58	73	perfect!

39. String **b** will give the same result.

Explanations will vary. Some students may try various numbers for the input to see if the output is the same. Others may realize that the order of the two numbers being added can be reversed but not the order of the addition and multiplication operations. String **a** is always 9 units less than the original.

Overview Students use arrow language and formulas to solve problems involving time.

About the Mathematics The formulas presented in this section were all written using complete words. They are sometimes called "word formulas" to distinguish them from formulas that use variables.

Do not encourage students to use letters to represent variables, unless they begin doing so by themselves. The intention of this unit is to give students a greater understanding of the structure and meaning of formulas.

Planning After students complete Section C, you may assign appropriate activities in the Try This! section, located on pages 54–58 of the Student Book, for homework.

Comments about the Problems

36. Informal Assessment This problem assesses students' ability to use a formula to solve a problem.

37. Informal Assessment This problem assesses students' ability to reason from a series of calculations to an informal formula.

Discuss the various formulas students create. Help them to realize that there is more than one possible formula that can be used to describe most situations. Students may start to use letters in their formulas as abbreviations for the miles and minutes. If they do so, have them write down what each letter stands for.

38. Informal Assessment This problem assesses students' ability to use reverse operations to find the input for a given output.

39. Informal Assessment This problem assesses students' ability to use and interpret formulas.

Section Focus

Students use reverse arrow strings to solve problems involving exchange rates and other contexts. The instructional focus of Section D is to:

- Use reverse arrow strings to solve problems.
- Use arrow string formulas to represent calculations with decimals and ratios.

Planning Instruction

Day 10. Foreign Money		Student pages 27–29
INTRODUCTION	Problems 1 and 2	■ Use an exchange rate to convert U.S. dollars into Dutch guilders.
CLASSWORK	Problems 3–10	■ Use arrow string formulas to convert U.S. dollars into Dutch guilders and vice versa.
HOMEWORK	Problems 11 and 12	■ Convert U.S. prices in dollars into Dutch guilders.

Additional Resources: Extension, page 71

Day 11. Going Backwards		Student pages 30–32
INTRODUCTION	Problems 13 and 14	■ Discuss a game involving reverse arrow strings.
CLASSWORK	Problems 15–19	■ Use reverse arrow strings to calculate the age of a tree.
HOMEWORK	Problems 20–22	■ Use arrow strings to determine the cost of deli items.

Additional Resources: Extension, page 73; Writing Opportunity, page 77; *Number Tools, Volume 2*, page 12

Day 12. Summary		Student page 32
INTRODUCTION	Review homework	■ Review homework from Day 11.
ASSESSMENT	Problems 23 and 24	■ Summary Questions.

Additional Resources: Try This! Section D, Student page 57; *Number Tools, Volume 2*, page 14

Materials

Student Resources	**Teacher Resources**	**Student Materials**
No resources required.	No resources required.	No materials required.

*See Hints and Comments for optional materials.

Concept Development

Reverse Arrow Strings to Solve Problems

Section D introduces the use of reverse arrow strings to solve problems involving exchange rates.

On **Day 10,** students investigate an exchange rate to convert U.S. dollars into Dutch guilders and vice versa. They use arrow string formulas that involve multiplication and division to represent and perform calculations with decimals and ratios.

On **Day 11,** students discuss a game involving reverse arrow strings. Given the arrow string and the output (the answer), they identify a strategy for finding the input (the starting number) of the formula. Next, students use reverse arrow strings to calculate the age of a tree. For homework, students use arrow strings to determine the cost of deli items.

On **Day 12,** students review homework from Day 11 and demonstrate their ability to reverse arrow strings and their understanding of the usefulness of reverse strings.

Planning Assessment

Problem 7
- Use and interpret simple formulas.
- Use reverse operations to find the input for a given output.
- Solve problems using the relationship between a mathematical procedure and its inverse.

Problem 14
- Use reverse operations to find the input for a given output.
- Solve problems using the relationship between a mathematical procedure and its inverse.

D. REVERSE OPERATIONS

FOREIGN MONEY

Marty is going to visit The Netherlands and has to buy Dutch guilders. The exchange rate is 1.65 Dutch guilders for every one United States dollar.

This arrow string shows the calculation:

$$\text{number of dollars} \xrightarrow{\times 1.65} \text{number of guilders}$$

1. How many Dutch guilders would Marty get for 50 United States dollars?

2. Copy and complete this table. Then explain how you found the numbers.

United States Dollars	1	2	3	4	5	6	7	8	9	10
Dutch Guilders										

3. In The Netherlands, Marty pays 2.45 Dutch guilders for a hamburger. Use the table to estimate about how much that would be in United States dollars.

4. a. Look closely at your table. Find a column in which the number of Dutch guilders is close to a whole number.

b. How could this discovery be useful?

1. 82.50 guilders. Students may use repeated addition, a calculator, or another number sense strategy to compute 50×1.65. Sample strategy using number sense:

   $50 = 10 \times 5$, so $50 \times 1.65 = 10 \times 1.65 + 5 \times 1.65 = 82.50$

   $10 \times 1.65 = 16.50$, and $5 \times 16.50 = 82.50$

2.

U.S. Dollars	1	2	3	4	5
Dutch Guilders	1.65	3.30	4.95	6.60	8.25

U.S. Dollars	6	7	8	9	10
Dutch Guilders	9.90	11.55	13.20	14.85	16.50

   Explanations will vary.

   Students may use a "double and add" approach (going from \$1 to \$2 to \$4 to \$8 by doubling, then finding combinations to add).

3. about \$1.50; 2.45 guilders is about halfway between \$1 and \$2.

4. **a.** Answers will vary. Some students might say that \$3 is close to 5 guilders. Others may respond that \$6 is close to 10 guilders.

   b. Answers will vary. Some students may say that these numbers (\$3 ≈ 5 guilders) are useful to make estimates when converting between U.S. dollars and Dutch guilders.

Materials foreign currency exchange rate table, optional (one per class); transparency of the table, optional (one per class); overhead projector, optional (one per class); calculators, optional (one per student)

Overview Students investigate how to exchange U.S. dollars for Dutch guilders.

About the Mathematics As in previous sections, students can use a variety of strategies other than the multiplication algorithm, since multiplying by 1.65 may be difficult for some students. You might suggest that they convert the dollar amounts to cents and then multiply to convert cents to guilders. Some students may use the strategy of repeated addition.

The conversion table that students complete in problem **2** (also known as a ratio table) may also be useful for the calculations to be done in these problems and those on the next page. Students may also use calculators to convert between the two monetary systems.

Planning You may want to start this section with a brief introduction about different currency systems.

Comments about the Problems

1. You might compare a current exchange rate from a local newspaper with the rate used here.

2. Some students may need to review strategies for using a ratio table. The ratio table is introduced in the units *Some of the Parts* and *Per Sense*.

3. You may need to help students realize that they must use reverse operations to solve this problem, since they are given the output and are asked to find the input. Have students check their estimate by using the formula and a calculator.

4. Some students may struggle with determining which decimal numbers are close to a whole number. You might have students who understand this concept explain their strategies to the rest of the class.

Did You Know? The exchange rate in this unit is 1.65 Dutch guilders for 1.00 U.S. dollar. That was the exchange rate in 1995. In 1992, however, the exchange rate was 2.00 Dutch guilders for 1.00 U.S. dollar. This shows that exchange rates often fluctuate.

To estimate prices in dollars so that he has an idea of how much he is paying, Marty uses the fact that 10 guilders is about 6 dollars. He thinks of the following rule:

$$\text{number of guilders} \xrightarrow{\div 10} \underline{\quad\quad} \xrightarrow{\times 6} \text{number of dollars}$$

5. Will this rule work? Explain your answer.

6. The prices for the items pictured below are given in Dutch guilders. Use the rule above to find the approximate cost of each item in United States dollars.

Nanda, Marty's Dutch friend, is coming to the United States for a visit. She wants to change Marty's conversion rule so that she can use it to find out how much things cost in guilders when the price is in dollars.

5. Yes. Students may give examples to show that this conversion rule produces results close to the actual amounts. Sample responses:

You divide by 10 to see how many groups of 10 guilders there are. Then each one is 6 dollars, so multiply the number of groups by 6.

$$13.20 \xrightarrow{\div 10} 1.32 \xrightarrow{\times 6} \$7.92$$

$$8.25 \xrightarrow{\div 10} 0.825 \xrightarrow{\times 6} \$4.95$$

6. Estimates will vary. Sample response:

Item	Guilder Amount	Dollar Amount
film	10	$6
flashbulbs	$2.98 \approx 3$	$1.80
dog food	$1.29 \approx 1.30$	$0.78
television	$499 \approx 500$	$300
entertainment center	$899 \approx 900$	$540

Materials calculators, optional (one per student)

Overview Students estimate to determine the dollar value of an item whose price is given in Dutch guilders.

About the Mathematics An exact formula is not always easy to use when making mental calculations. It can be helpful to use a simpler rule, although the result may not be as exact as that in the original formula. For example, the arrow string shown on Student Book page 28 can be shortened to a string involving only one calculation: $\times \frac{6}{10}$ or $\times 0.6$. Although some students may come up with this shortcut on their own, do not make this explicit at this point in the unit.

Planning After students have finished problem **6,** you may want to have a class discussion about the two different rules students used to convert between U.S. dollars and guilders. You may also want to discuss the estimation strategies that students may have used.

Comments about the Problems

5–6. These problems are critical. Be sure that students understand how to estimate before they begin solving these two problems. Suggest that they look carefully on the table on Student Book page 27. They may need to think about how they could use the $6 to 10 guilders relationship to convert $12 or $24 to guilders. You may ask students to think about how their estimations would change as the dollar amount grows larger [for each multiple of $3, add another 5 guilders].

6. You might suggest that students first round the amounts in guilders before computing.

Marty's rule:

number of guilders $\xrightarrow{\div 10}$ _____ $\xrightarrow{\times 6}$ number of dollars

7. a. How can Marty's rule be changed to convert United States dollars to Dutch guilders? Write the new rule as an arrow string.

   b. Does your rule from part **a** agree with the exchange rate given at the beginning of this section?

Marty is surprised that the new rule seems so much harder than the one he used.

8. Why is the rule for converting United States dollars to Dutch guilders harder?

Nanda thinks of a simpler rule:

number of dollars $\xrightarrow{\times 3}$ _____ $\xrightarrow{\div 2}$ number of guilders

She also thinks of another rule:

number of dollars $\xrightarrow{\times 5}$ _____ $\xrightarrow{\div 3}$ number of guilders

9. How well do these rules work?

10. If you were Nanda, which rule would you use? Why?

11. Think of three things that someone visiting your city might want to buy. For each item, estimate the price in United States dollars. Then calculate how much it would cost in Dutch guilders. Use the exchange rate given in this section.

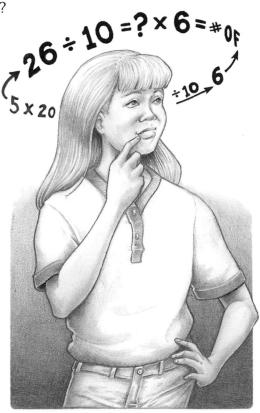

Lisa and Andre are pen pals. In a letter to Andre, Lisa wrote that she paid $5 to see a movie in the United States. Andre had seen the same movie in The Netherlands and paid 13 Dutch guilders.

12. In which country did the movie cost more? Show how you found your answer.

7. a. dollars $\xrightarrow{\div 6}$ ____ $\xrightarrow{\times 10}$ guilders

b. It is very close to the original exchange rate. Students can determine this by using a calculator to multiply dollars by 10 and divide the result by 6 to check several of the amounts entered in the table for problem **2**.

8. Answers will vary. Sample responses:

The rule is harder because now I have to divide by six rather than by 10, and dividing by six is harder than dividing by 10.

9. Answers will vary. Sample responses:

The first rule is not very exact and the estimate is a little too small. The second rule is more exact.

The first rule is easy to apply and it gives a reasonable estimate. The second rule is harder to apply, though it is more accurate.

10. Answers will vary. Some students may like the rule in which you divide by two because that is easy for them to do mentally. Others may choose one of the other rules because it is more accurate.

11. Answers will vary. Sample responses:

A $10 T-shirt will cost 10 × 1.65 = 16.50 guilders.

A $25 sweater will cost 25 × 1.65 = 41.25 guilders.

A $5 book will cost half of 16.50 (the $10 T-shirt), or 8.25 guilders.

12. It costs more to see a movie in the Netherlands (13 guilders) than it does in the United States (8.25 guilders). Strategies will vary. Sample strategy:

$5 x 1.65 = 8.25 guilders in the U.S. compared to 13 guilders in the Netherlands.

OR

13 ÷ 1.65 = $7.87 U.S. dollars in the Netherlands compared to $5 in the U.S.

Materials calculators, optional (one per student)

Overview Students use other estimate rules to convert guilder amounts into dollar amounts and reverse the calculation to convert dollars to guilders.

Planning Students should have a good understanding of how to reverse arrow strings and be able to apply it to the problems on this page. If some students are still using a guess and check method, you might suggest that it would be more efficient for them to have a rule that can be used quickly to find the answer.

Comments about the Problems

7. Informal Assessment This problem assesses students' ability to use and interpret simple formulas, use reverse operations to find the input for a given output, and to solve problems using the relationship between a mathematical procedure and its inverse.

Students should realize that they need to reverse the arrow string to answer part **a**. Students who have not yet mastered this may benefit from seeing an arrow string side-by-side with its reverse arrow string and comparing the two strings.

8. Some students may find division more difficult than multiplication.

11. Encourage students to be creative in the objects they choose. Have students explain how they found the amount in guilders for each object and discuss why some of the strategies are reasonable. You might pair students up and have one list the objects and their prices in dollars and the other find the corresponding price in guilders, then switch their roles.

Extension Change the numbers in problem **12** to $5.00 and 8 guilders. Students will discover that, although 8 is more than 5, after converting the amounts, the dollar amount is more. Ask students to find out for what guilder amounts the Dutch movies would be less expensive than the U.S. movies and for what amounts they would be more expensive. [Five dollars is the same as 8.25 guilders, so when the Dutch movies cost less than 8.25 the U.S. movies cost more; when the Dutch movies cost more than 8.25, the U.S. movies cost less.]

GOING BACKWARDS

Pat and Kris are playing a game. One player writes down an arrow string and the output (answer) but not the input (starting number). The other player has to determine the input.

Here is Pat's arrow string and output:

$$\underline{\quad?\quad}\xrightarrow{+4}\underline{\quad\quad}\xrightarrow{\times 10}\underline{\quad\quad}\xrightarrow{-2}\underline{\quad\quad}\xrightarrow{\div 2}29$$

13. a. What should Kris give as the input? Explain how you came up with this number.

  b. One student found an answer for Kris by using a *reverse string*. What should go above each of the reversed arrows below?

$$\underline{\quad?\quad}\longleftarrow\underline{\quad\quad}\longleftarrow\underline{\quad\quad}\longleftarrow\underline{\quad\quad}\longleftarrow 29$$

On the next round, Kris wrote:

$$\underline{\quad?\quad}\xrightarrow{+3}\underline{\quad\quad}\xrightarrow{\div 6}\underline{\quad\quad}\xrightarrow{+5}\underline{\quad\quad}\xrightarrow{-2}6$$

14. a. What should Pat give as the input? Explain how you found this number.

  b. Write the reverse string that can be used to find the input.

13 **a.** The input number is 2. Strategies will vary. Sample strategy:

Some students may work backwards, using inverse operations in the arrow string:

$$29 \xrightarrow{\times 2} 58 \xrightarrow{+2} 60 \xrightarrow{\div 10} 6 \xrightarrow{-4} 2$$

b. $\times 2, + 2, \div 10, - 4$ (from right to left).

The arrow string should look like this:

$$? \xleftarrow{-4} \underline{\quad} \xleftarrow{\div 10} \underline{\quad} \xleftarrow{+2} \underline{\quad} \xleftarrow{\times 2} 29$$

14. a. The input number is 15. Explanations will vary, but students may use arrow language or working backwards.

b. $15 \xleftarrow{-3} 18 \xleftarrow{\times 6} 3 \xleftarrow{-5} 8 \xleftarrow{+2} 6$

Overview Students read about a game in which an arrow string with its output is given and the input must be determined. Students use reverse arrow strings and inverse operations to find the input.

Comments about the Problems

14. Informal Assessment This problem assesses students' ability to use reverse operations to find the input for a given output and solve problems using the relationship between a mathematical procedure and its inverse.

Extension Ask students to create long arrow strings in which the input is the same as the output. The arrow strings need to be at least four steps long. (This implies that each operation in the string needs to be undone by another operation in the string.)

Beech Trees

There are some beech trees in a park near Jessica's house. Over the years, a number of botanists have studied these trees very carefully. They discovered something very interesting. When a tree is between 20 and 80 years old, it grows fairly evenly. They developed two formulas that describe the growth of the trees if the age is known.

age $\xrightarrow{\times 0.4}$ ___ $\xrightarrow{-2.5}$ thickness   age $\xrightarrow{\times 0.4}$ ___ $\xrightarrow{+1}$ height

For these formulas, the age is in years; the height is in meters; the thickness is in centimeters and is measured 1 meter from the ground.

15. Find the heights and thicknesses of trees that are 20, 30, and 40 years old.

16. Jessica is interested in knowing the age of a tree. How can she find it?

Thickness refers to the diameter of the tree trunk. Jessica uses some straight sticks to help her measure the thickness of a tree. She finds that it is 25.5 centimeters.

17. How old is the tree?

Jessica estimates the height of another beech tree to be about 20 meters.

18. Use this estimate of the height to calculate the age of the tree.

Jessica realizes that she can make a new formula. Her new formula gives the height of a tree if the thickness is known.

19. Write Jessica's formula.

15. A 20-year-old tree will be 9 meters tall and 5.5 centimeters thick;

a 30-year-old tree will be 13 meters tall and 9.5 centimeters thick;

a 40-year-old tree will be 17 meters tall and 13.5 centimeters thick.

Sample strategy: Using arrow language:

$$\text{age} \xrightarrow{\times 0.4} \underline{\quad} \xrightarrow{-2.5} \text{thickness}$$
$$20 \xrightarrow{\times 0.4} 8 \xrightarrow{-2.5} 5.5$$
$$30 \xrightarrow{\times 0.4} 12 \xrightarrow{-2.5} 9.5$$
$$40 \xrightarrow{\times 0.4} 16 \xrightarrow{-2.5} 13.5$$

$$\text{age} \xrightarrow{\times 0.4} \underline{\quad} \xrightarrow{+1} \text{height}$$
$$20 \xrightarrow{\times 0.4} 8 \xrightarrow{+1} 9$$
$$30 \xrightarrow{\times 0.4} 12 \xrightarrow{+1} 13$$
$$40 \xrightarrow{\times 0.4} 16 \xrightarrow{+1} 17$$

16. Answers will vary.

She could estimate either the height or thickness, then reverse the arrow string to find the age.

17. The tree is 70 years old. Most students will use reverse arrow strings to solve this problem.

$$25.5 \xrightarrow{+2.5} 28 \xrightarrow{\div 0.4} 70$$

18. The tree is about 47.5 years old. Most students will use reverse arrow strings to solve this problem.

$$20 \xrightarrow{-1} 19 \xrightarrow{\div 0.4} 47.5$$

19. $$\text{thickness} \xrightarrow{+2.5} \underline{\quad} \xrightarrow{\div 0.4} \text{age} \xrightarrow{\times 0.4} \underline{\quad} \xrightarrow{+1} \text{height}$$

Some students may realize they can shorten this string to:

$$\text{thickness} \xrightarrow{+3.5} \text{height}$$

Materials calculators, optional (one per student)

Overview Students investigate two formulas that describe the growth of beech trees when the age of the tree is known.

About the Mathematics Formulas can be combined, as in problem **19,** when the output of one formula is used as the input for the other. By using a reverse arrow string on one of the formulas, students can connect height directly to thickness. Combining formulas (composition of functions) is further investigated in the last section of this unit and in the unit *Building Formulas.*

Planning Problem **19** may be difficult for some students as it requires abstract thinking to shorten an arrow string. Students may work in small groups on these problems.

Comments about the Problems

15. Students use two different formulas to find the age of a tree, depending on what information they are given. You may want to mention that the answers are approximate because tree growth depends heavily on climate and environmental conditions.

17. You may want to have students use a calculator to divide and multiply by 0.4. Otherwise, dividing by 0.4 can be simplified by first multiplying by 10 and then dividing by 4.

18. If students are concerned that the same tree has two different ages, you can point out that the ages are close and that an error in measuring could account for the difference.

Did You Know? A more common way of finding the age of a tree is cutting through the tree and counting the age rings in its trunk. In the unit *Ups and Downs,* this method is further investigated and related to graphs to describe the growth of trees.

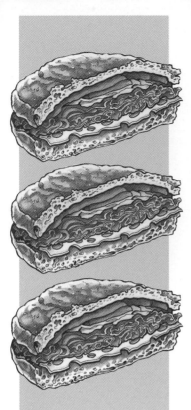

Ham and Cheese

Carmen and Andy are at the store buying ham and cheese for sandwiches. Carmen is going to buy Swiss cheese that costs $4.40 per pound. She decides that she wants to buy 0.75 pound. She wants to know how much it will cost before she orders it.

20. Write an arrow string to show how much it will cost.

This is the string that Carmen wrote:

$$\$4.40 \xrightarrow{\div 4} \$1.10 \xrightarrow{\times 3} \$3.30$$

21. Is Carmen's string correct? What did she do?

Carmen meets Andy, who has already bought some ham. Andy bought 0.75 pound, and it cost $2.25. Carmen wonders what the price per pound is.

22. Write an arrow string to find the price per pound of ham.

Summary

Every arrow has a reverse arrow. A reverse arrow has the opposite operation.

For example, the reverse of $\xrightarrow{\div 4}$ is $\xleftarrow{\times 4}$.

Reverse arrows can be used to make reverse strings. For example,

____ $\xrightarrow{\div 4}$ ____ $\xrightarrow{\times 3}$ ____ reverses to ____ $\xleftarrow{\times 4}$ ____ $\xleftarrow{\div 3}$ ____ ,

which is the same as ____ $\xrightarrow{\div 3}$ ____ $\xrightarrow{\times 4}$ ____ .

Summary Questions

23. When are reverse strings useful?

24. Write the reverse for each of the following strings:

    a. input $\xrightarrow{+ 2}$ ____ $\xrightarrow{\times 3}$ ____ $\xrightarrow{- 4}$ output

    b. input $\xrightarrow{\div 2}$ ____ $\xrightarrow{- 5}$ ____ $\xrightarrow{+ 7}$ output

    c. Use some numbers to find out whether or not your reverse strings are correct.

20. $4.40 \xrightarrow{\times \, 0.75 \text{ lb}} \$3.30$

21. Yes. Carmen's string is correct because 0.75 is the same as $\frac{3}{4}$ which is the same as $3 \div 4$. So, first she calculated one-fourth of the price. Next, she multiplied by three which gives three-fourths of the price.

22. Ham costs $3 per pound. Strategies will vary. Sample strategies:

Strategy 1

Some students may use a reverse arrow string using 0.75.

$2.25 \xrightarrow{\div \, 0.75} \$3$

Strategy 2

Some students may use a different reverse arrow string:

$2.25 \xrightarrow{\div \, 3} \$0.75 \xrightarrow{\times \, 4} \$3$

23. Reverse arrow strings are useful when you know the answer (output) and have to find what the original number (input) is.

24. a. input $\xleftarrow{-\,2}$ _____ $\xleftarrow{\div \, 3}$ _____ $\xleftarrow{+\,4}$ output

   b. input $\xleftarrow{\times \, 2}$ _____ $\xleftarrow{+\,5}$ _____ $\xleftarrow{-\,7}$ output

   c. Students should use several numbers to check to see if their arrow strings work.

Overview Students practice writing arrow strings and reverse arrow strings.

About the Mathematics This page reviews arrow strings and reverse arrow strings. In problems **20–22,** fractions, decimals, and operations are related to each other. In the unit *Fraction Times*, the connections among fractions, percents, and decimals are made explicit. In *Expressions and Formulas*, it is assumed that students know how these rational numbers relate to each other.

Planning You may want to have a class discussion comparing the benefits of reversing arrow strings vs. using the guess and check strategy. The problems on this page provide practice in converting between decimals and fractions. After students complete Section D, you may assign appropriate activities in the Try This! section, located on pages 54–58 of the Student Book, for homework.

Comments about the Problems

   21. Some students may need to review the relationships between the given fraction and decimal to solve this problem. If necessary, refer to *Number Tools* Volumes 1 and 2 for additional work.

Writing Opportunity Have students write about arrow strings, using problems **23** and **24** as examples in their paragraphs. You might collect their work to help you evaluate students' understanding of arrow language and whether or not they can work with reverse arrow strings.

Section Focus

Students use reverse arrow strings to solve problems involving exchange rates and other contexts. The instructional focus of Section E is to:

* Use reverse arrow strings to solve problems.
* Use arrow string formulas to represent calculations with decimals and ratios.

Planning Instruction

Day 13. Apple Crisp		Student pages 33–37
INTRODUCTION	Problem 1	■ Use a recipe to determine the amount of ingredients needed for different serving sizes.
CLASSWORK	Problems 4–9	■ Use arrow strings to describe the numbers in a bus schedule table.
HOMEWORK	Problems 2, 3, 10–12	■ Solve problems involving tables and calculation charts.

Additional Resources: Bringing Math Home, page 81; Extension, page 83

Day 14. Home Repairs		Student pages 38–40
INTRODUCTION	Problem 13	■ Calculate the cost of plumbing repairs using addition and multiplication.
CLASSWORK	Problems 14–18	■ Use arrow strings to describe regularities in a table and extend the table.
HOMEWORK	Problems 19 and 20	■ Find a table in a newspaper that has regularities and one that does not.

Additional Resources: Writing Opportunity, page 95; Try This! Section E, Student page 57; *Number Tools, Volume 2,* page 16

Materials

Student Resources

Quantities listed are per student.

- Student Activity Sheets 1–4

Teacher Resources

No resources required.

Student Materials

No materials required.

*See Hints and Comments for optional materials.

Concept Development

Identify Regularities in a Table Using Arrow Strings

Section E introduces tables that have patterns that can be generated or described using *arrow string formulas.*

On **Day 13,** students use a recipe to determine the amount of ingredients needed for different serving sizes. (Note: This problem introduces the context for two of the homework problems.) Next, students use arrow strings to describe the patterns in a bus schedule and to solve problems involving other tables.

On **Day 14,** students calculate the cost of plumbing repairs that include travel expenses and per-hour rate. The calculations involve both addition and multiplication. The context of the problem helps students to keep the operations in the correct order. After calculating the cost of several repair jobs, students use arrow strings to describe patterns in a cost-of-work table. For homework, they find a table in a newspaper that has regularities and one that does not.

Planning Assessment

Problems 9, 11, 15
- Interpret relationships displayed in tables.

Problem 16
- Interpret relationships displayed in tables.
- Use and interpret simple formulas.
- Generalize from patterns to symbolic relationships.
- Use formulas in any representation (arrow language, arithmetic trees, words) to solve problems.

E. TABLES

Apple Crisp

APPLE CRISP–*serves four*

You will need:

250 grams of flour

125 grams of butter

1 egg

450 grams of apples

water, salt, cinnamon, lemon juice, and sugar

The chef at the Very Good Restaurant bakes apple crisp using a recipe in the Very Good Cookbook. The size of the apple crisp is determined by the number of people it must serve.

The chef has to calculate the amounts of ingredients needed for different sizes of apple crisp. The original recipe makes an apple crisp that will serve four people. One day, the chef receives an order for an apple crisp to serve six people. The chef begins the calculations by finding the amounts of ingredients needed to serve two people.

Ingredients	4 servings	2 servings	6 servings
Flour	250 g		
Butter	125 g		
Eggs	1		
Apples	450 g		

Use **Student Activity Sheet 1** to answer problems **1–3**.

1. a. Fill in the missing values in the first table on **Student Activity Sheet 1.**

b. Why do you think that the chef started with a two-serving apple crisp?

Ingredients	Servings							
	2	3	4	5	6	8	10	12
Flour			250 g					
Butter			125 g					
Eggs			1					
Apples			450 g					

2. Fill in the second table on **Student Activity Sheet 1.**

3. Describe some of the strategies you used to fill in the table.

1. a.

Ingredients	4 servings	2 servings	6 servings
flour	250 g	125 g	375 g
butter	125 g	62.5 g	187.5 g
eggs	1	0.5	1.5
apples	450 g	225 g	675 g

b. Answers will vary, but knowing the amounts for two servings of apple crisp helps in determining the amounts needed for six servings. You can either double this amount or add the four servings with the two servings.

2.

Ingredients	Servings							
	2	3	4	5	6	8	10	12
Flour	125	187.5	250	312.5	375	500	625	750
Butter	62.5	93.75	125	156.25	187.5	250	312.5	375
Eggs	$\frac{1}{2}$	$\frac{3}{4}$	1	$1\frac{1}{4}$	$1\frac{1}{2}$	2	$2\frac{1}{2}$	3
Apples	225	337.5	450	562.5	675	900	1125	1350

Gram amounts in the table may be rounded to the nearest gram or half gram.

3. Strategies will vary. Sample strategies:

The ingredient amount for two servings is one-half of those used in four servings (See table in **2.**)

The ingredient amount for six servings is three times those used in two servings, or four servings added to two servings.

The ingredient amount for three servings is one-half of those used in six servings.

The ingredient amount for five servings is one-half of those used in 10 servings, or three servings added to two servings.

The ingredient amount for eight servings is two times those used in four servings, or six servings added to two servings.

The ingredient amount for 10 servings is two times those used in five servings, or six servings added to four servings.

The ingredient amount for 12 servings is two times those used in six servings, or eight servings added to four servings.

Materials Student Activity Sheet 1 (one per student); copies of apple crisp recipe shown below in the Bringing Math Home activity, optional (one per student)

Overview Students calculate amounts of ingredients needed for different-size servings of apple crisp.

About the Mathematics The tables on this page are ratio tables, as introduced in *Some of the Parts*. Ratio tables are also used in *Per Sense* and *Grasping Sizes*, and many other *Mathematics in Context* units in the number strand.

Planning Even students with no previous experience with ratio tables should be able to complete the tables. A discussion about problem **1** should include how a student could actually measure only $\frac{1}{2}$ of an egg for the 2-serving recipe. Some students may suggest using 1 egg for this recipe. Others may suggest using an egg substitute that can be measured or beating the egg and halving the mixture.

Comments about the Problems

1–3. As shown in the solutions column, students may use a variety of strategies to complete the tables. Encourage students to share their strategies in class.

Bringing Math Home Some students may want to make this microwave apple crisp recipe at home with adult supervision:

Ingredients:

$1\frac{1}{2}$ cups peeled, sliced, and coarsely chopped tart apples

2 tablespoons flour

2 tablespoons quick-cooking oats

2 tablespoons packed brown sugar

2 tablespoons margarine or butter, softened

$\frac{1}{8}$ teaspoon ground cinnamon

$\frac{1}{8}$ teaspoon ground nutmeg

Spread apples in a 24-ounce, microwave-safe casserole dish. Mix remaining ingredients until crumbly and spread mixture over apples. Place uncovered dish inside microwave oven. Cook for about 5 minutes on high power or until apples are tender. Serve warm with cinnamon or ice cream. Makes two servings.

Shuttle Service

The Lake Shore Bus Company runs a shuttle service between Chicago and northwest Indiana. Steve has a copy of the timetable for the service.

O'Hare International Airport ● **Expo Center** ⑨⓪ ⑨④

The bus company decides to make a new stop at the Expo Center.

O'HARE TO NORTHWEST INDIANA

Find the time you leave O'Hare in the left column.
Read straight across on the same line to your destination point.
This will show your arrival time at that point.

Leave O'Hare Lower Terminal	Arrive Expo Center	Arrive Hammond/ Highland	Arrive Glen Park	Arrive Merrillville
5:45 A.M.		7:05 A.M.	7:20 A.M.	7:40 A.M.
6:45 A.M.		8:05 A.M.	8:20 A.M.	8:40 A.M.
7:45 A.M.		9:05 A.M.	9:20 A.M.	9:40 A.M.
8:45 A.M.		10:05 A.M.	10:20 A.M.	10:40 A.M.
9:45 A.M.		11:05 A.M.	11:20 A.M.	11:40 A.M.
10:45 A.M.		12:05 P.M.	12:20 P.M.	12:40 P.M.
11:45 A.M.		1:05 P.M.	1:20 P.M.	1:40 P.M.
12:45 P.M.		2:05 P.M.	2:20 P.M.	2:40 P.M.
1:45 P.M.		3:05 P.M.	3:20 P.M.	3:40 P.M.
2:45 P.M.		4:05 P.M.	4:20 P.M.	4:40 P.M.

Steve's job is to add the new column for the Expo Center stop. Luckily, he knows that the bus arrives at the Expo Center 15 minutes after it leaves O'Hare's Lower Terminal.

4. Use the timetable on **Student Activity Sheet 2** to fill in the times for the Expo Center stop.

4.

O'HARE TO NORTHWEST INDIANA

Reading schedules from O'Hare:
Find the time you leave O'Hare in the left column. Read straight across to the right on the same line to your destination point. This will show your arrival time at that point.

Leave O'Hare Lower Terminal	Arrive Expo Center	Arrive Hammond/ Highland	Arrive Glen Park	Arrive Merrillville
5:45 A.M.	6:00 A.M.	7:05 A.M.	7:20 A.M.	7:40 A.M.
6:45 A.M.	7:00 A.M.	8:05 A.M.	8:20 A.M.	8:40 A.M.
7:45 A.M.	8:00 A.M.	9:05 A.M.	9:20 A.M.	9:40 A.M.
8:45 A.M.	9:00 A.M.	10:05 A.M.	10:20 A.M.	10:40 A.M.
9:45 A.M.	10:00 A.M.	11:05 A.M.	11:20 A.M.	11:40 A.M.
10:45 A.M.	11:00 A.M.	12:05 P.M.	12:20 P.M.	12:40 P.M.
11:45 A.M.	12:00 P.M.	1:05 P.M.	1:20 P.M.	1:40 P.M.
12:45 P.M.	1:00 P.M.	2:05 P.M.	2:20 P.M.	2:40 P.M.
1:45 P.M.	2:00 P.M.	3:05 P.M.	3:20 P.M.	3:40 P.M.
2:45 P.M.	3:00 P.M.	4:05 P.M.	4:20 P.M.	4:40 P.M.

Strategies will vary. Some students may add 15 minutes to each entry in column 1.

Others may begin by adding 15 minutes to the first entry in column 1. Then, since each entry in column 1 is one hour apart, students may fill in column 2 by adding one hour to the first entry to get 7:00 A.M., and so on.

Materials Student Activity Sheet 2 (one per student); bus, subway, or train schedules, optional (one per student)

Overview Students investigate patterns in a timetable of an airport shuttle service. They expand the schedule to include times for a new destination along the route. This context is continued on the next page.

About the Mathematics Patterns in tables are represented here using both arrow strings and formulas. Students develop greater flexibility in moving between the different mathematical representations as they progress through the various algebra strand units. They also gain a better understanding of the similarities and differences between the different representations: the real-life situation, table, graph, and formula.

Planning You may introduce this context with a brief class discussion about bus routes and timetables.

Comments about the Problems

4. Students can complete the table in at least two different ways. They may come up with their own strategies as well.

Extension You might bring in bus, subway, or train schedules and have students examine the different timetables for patterns. Some may have patterns that are not easily described by a formula. For example, more buses may run during certain times of the day. Students could also be asked to search for other tables that have patterns and those that do not. Examples of some that might have patterns are:

• charts for serving dog food to puppies of different sizes or ages,

• tables for making different amounts of instant oatmeal,

• tables describing age and target pulse rates during exercise, or

• charts showing how weight and height relate.

You may have noticed that the second column can be filled in by adding 15 minutes to the times in the first column. This can be shown with an arrow string.

$$\text{Lower Terminal} \xrightarrow{\ +\ 15\ \text{minutes}\ } \text{Expo Center}$$

5. The other columns can be created in much the same way. Copy and complete the string below.

$$\text{Lower Terminal} \xrightarrow{\ +\ 15\ \text{minutes}\ } \text{Expo Center} \xrightarrow{\ ?\ } \text{Hammond} \xrightarrow{\ ?\ } \text{Glen Park} \xrightarrow{\ ?\ } \text{Merrillville}$$

6. a. There is also a pattern going down each column. How can this regularity be explained?

b. Use the pattern to add an extra row to the timetable on **Student Activity Sheet 2.**

The timetable for the Lake Shore Bus Company can be generated with two arrow strings: a horizontal arrow string and a vertical arrow string.

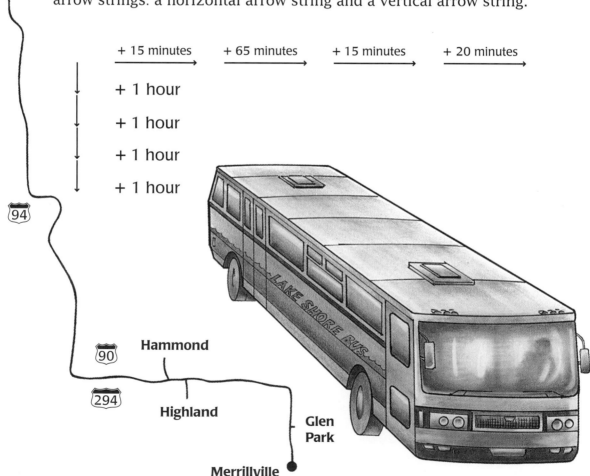

5. Terminal $\xrightarrow[\text{minutes}]{+\,15}$ Expo Center $\xrightarrow[\text{5 minutes}]{+\,1\text{ hour and}}$

   Hammond $\xrightarrow[\text{minutes}]{+\,15}$ Glen Park $\xrightarrow[\text{minutes}]{+\,20}$ Merrillville

6. **a.** Each row shows a time one hour later than the previous row because the bus leaves every hour.

   b. The last row would be

   3:45 P.M. 4:00 P.M. 5:05 P.M. 5:20 P.M. 5:40 P.M

Materials Student Activity Sheet 2 (one per student)

Overview Students use arrow strings to describe patterns in the timetable introduced in problem **4**.

Comments about the Problems

5. The arrow string can also be shortened to one arrow by finding the sum of all the times between stops. The total time it takes to travel from O'Hare to Merrillville is 1 hour and 55 minutes.

6. Encourage students to look for patterns in the table. Some students may realize that they can merely add one hour to each of the entries in the last row in the timetable to generate the new row.

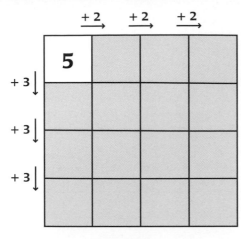

Like the timetable for the bus, other tables can be created from a pair of strings.

7. Complete the table at the top of **Student Activity Sheet 3.**

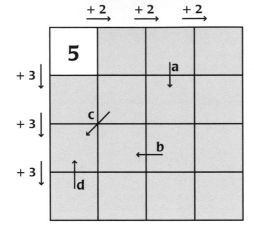

8. What operations fit with the arrows marked **a** through **d** on the table in the middle of **Student Activity Sheet 3?**

Instead of having an arrow for each row and column, a table can be written with one arrow to show the change for every row, and one arrow to show the change for every column.

The table on the left can be filled in by multiplying by five for each move to the right and by two for each move down.

9. Which operations fit with the arrows marked **a** through **c** inside the table?

7.

	$\xrightarrow{+2}$	$\xrightarrow{+2}$	$\xrightarrow{+2}$
5	7	9	11
8	10	12	14
11	13	15	17
14	16	18	20

(with $+3 \downarrow$ between each row)

8.

arrow a	+ 3
arrow b	− 2
arrow c	+ 1
arrow d	− 3

9.

arrow a	× 25
arrow b	÷ 5
arrow c	÷ 2

Materials Student Activity Sheet 3 (one per student)

Overview Students investigate patterns in tables similar to the timetable for the bus, and use arrow string formulas to describe the operations used to fill in the table.

About the Mathematics The first two tables are generated by repeated addition. The first and second table can be filled in to answer question **8.** For question **9,** students are asked to only look at a table without entries and use the operations to determine what each arrow stands for. The tables on this and the next page are calculation charts. Similar charts are used in the unit *Comparing Quantities*, where entries represent prices of combinations of items.

Comments about the Problems

8. You may want to point out to students that the horizontal and vertical formulas are different. When students fill in the table, they will probably work from left to right and from top to bottom. To check their work, you may suggest that they start in the lower right corner and reverse the operations.

9. **Informal Assessment** This problem assesses students' ability to interpret relationships displayed in tables.

If students are having difficulty, have them work in pairs in which one student makes up a table with arrows that indicate the computation to be performed and the other student fills in the table. At this point you may suggest that the operation for one table be either addition and subtraction or multiplication and division. Combining addition with multiplication is addressed on the next page. You may want to point out that arrow **a** crosses two squares, so the rule is × 25, not × 5.

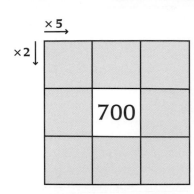

×5 →

×2 ↓

	700	

+4 →

×3 ↓

2		

10. Only one number in the table on the left is known. Write the missing numbers in the table at the bottom of **Student Activity Sheet 3.**

The table on the left is different from the previous ones.

11. Copy and complete the table.

12. Compare your answers with those of your classmates. Do some students have different numbers in the table? If so, why?

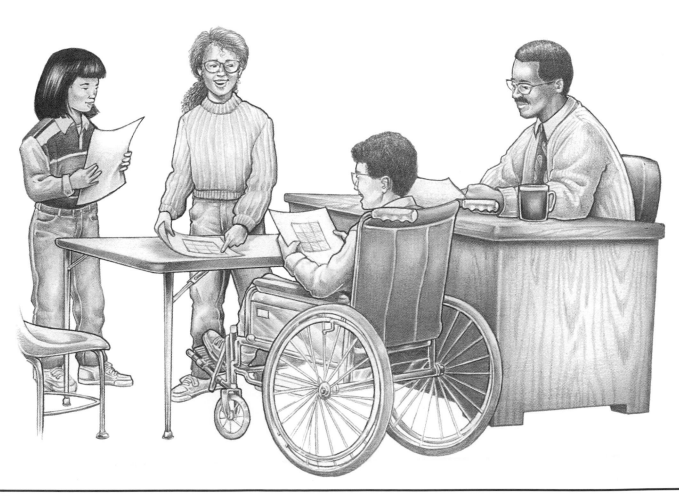

10.

×2 ↓	×5 →		
70	350	1750
140	700	3500
280	1400	7000

11. Answers will vary, depending on students' strategies in filling in the chart. Sample student responses and strategies:

Strategy 1

After filling in the top row and left column, some students may fill in the rows according to the + 4 arrow.

2	6	10
6	10	14
18	22	26

Strategy 2

After filling in the top row and left column, some students may fill in the columns according to the x 3 arrow.

2	6	10
6	18	30
18	54	90

Strategy 3

Some students may fill in the columns and rows going back and forth along the rows. Sample response:

2	6	10
22	26	30
66	70	74

12. Answers will vary. Many students will notice that their classmates' tables have different numbers, depending on the order in which the cells in each table were completed. For example, filling in the columns first gives a different solution than filling in the rows first. The reason that the answers are different is because this table involved both addition and multiplication operations. With these two operations, the order in which the calculations are done is important.

Materials Student Activity Sheet 3 (one per student)

Overview Students use arrow strings to complete two more calculation tables.

About the Mathematics The tables in problems **7** through **10** each involve only one operation, addition or multiplication. The table used in problem **11** involves both addition and multiplication. Because the order in which addition and multiplication are performed together affects the result, there are various solutions for completing this table. This complicates the situation for students and creates a need to have a mathematical connection for order of operations.

Comments about the Problems.

11. Informal Assessment This problem assesses students' ability to interpret relationships displayed in tables.

Allow students to complete the table on their own and compare their table with that of a classmate. You might ask them what problems they encountered.

This is the first real problem where the need for an order of operations is addressed. It is necessary to mention that—unless a definite order or pattern is predetermined—there will be many possible solutions. Some students may need to be shown that this table is different from the other tables because it combines addition and multiplication. This illustrates the need for a standard order of operations.

Home Repairs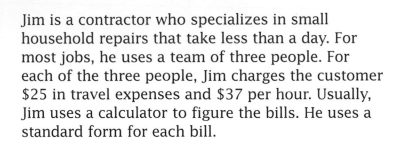

Jim is a contractor who specializes in small household repairs that take less than a day. For most jobs, he uses a team of three people. For each of the three people, Jim charges the customer $25 in travel expenses and $37 per hour. Usually, Jim uses a calculator to figure the bills. He uses a standard form for each bill.

Jim MacIntosh Total Repairs
147 Franklin Rd., Wakeshire
Customer: _____
Labor _____ hours at $37/hour . . . . $_____
Travel costs: . . . . $ 25.00
_____
Total cost per worker . . . . $_____
Total bill = total cost per worker × 3 . . . . $_____
(3 workers)

13. Use the forms on **Student Activity Sheet 4** to show what Jim would charge for each of the plumbing repair jobs shown on the right.

a. replacing pipes for Mr. Ashton: 3 hours

b. cleaning out the pipes at Rodriguez and Partners: $2\frac{1}{2}$ hours

c. replacing faucets at the Vander house: $\frac{3}{4}$ hour

13.

a.

> ## Jim MacIntosh Total Repairs
> ## 147 Franklin Rd., Wakeshire
>
> Customer: ___Mr. Ashton___
>
> Labor ___3___ hours at $37/hour . . . .$ _111.00_
>
> Travel costs: . . . .$ 25.00
>
> Total cost per worker . . . .$ _136.00_
>
> Total bill = total cost per worker × 3 . . . . . .$ _408.00_
>
> (3 workers)

b.

> ## Jim MacIntosh Total Repairs
> ## 147 Franklin Rd., Wakeshire
>
> Customer: ___Rodriguez and Partners___
>
> Labor ___2.5___ hours at $37/hour . . . .$ _92.50_
>
> Travel costs: . . . .$ 25.00
>
> Total cost per worker . . . .$ _117.50_
>
> Total bill = total cost per worker × 3 . . . . . .$ _352.50_
>
> (3 workers)

c.

> ## Jim MacIntosh Total Repairs
> ## 147 Franklin Rd., Wakeshire
>
> Customer: ___Vander___
>
> Labor $\frac{3}{4}$ hours at $37/hour . . . .$ _27.75_
>
> Travel costs: . . . .$ 25.00
>
> Total cost per worker . . . .$ _52.75_
>
> Total bill = total cost per worker × 3 . . . . . .$ _158.25_
>
> (3 workers)

Materials Student Activity Sheet 4 (one per student)

Overview Students use a combination of multiplication and addition computations to calculate the total costs for different home plumbing repair jobs.

About the Mathematics Students are required to interpret and organize information to solve problems involving plumbing bills for household repairs. Although these problems involve both addition and multiplication, the context of the problems and the format of the billing statement help students to keep the operations in order.

Comments about the Problems

13. Some students may use arrow language as illustrated in the solutions column. Other students may use another representation.

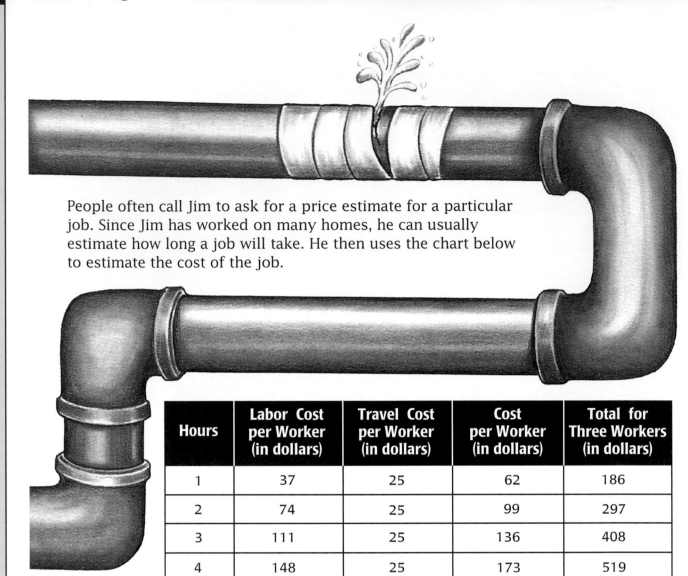

People often call Jim to ask for a price estimate for a particular job. Since Jim has worked on many homes, he can usually estimate how long a job will take. He then uses the chart below to estimate the cost of the job.

Hours	Labor Cost per Worker (in dollars)	Travel Cost per Worker (in dollars)	Cost per Worker (in dollars)	Total for Three Workers (in dollars)
1	37	25	62	186
2	74	25	99	297
3	111	25	136	408
4	148	25	173	519

14. a. What do the entries in the first row of the table mean?

b. What would the row for 5 hours look like?

15. a. Explain the regularity in the column for labor cost per worker.

b. Look carefully at the table. Make a list of all of the regularities you can find. Explain the regularities.

16. a. Draw an arrow string that Jim could use to make more rows for the table.

b. Use your arrow string to make two more rows (for 6 and 7 hours) for the table.

17. Can you use arrow strings to make any table? Explain.

14. a. The entries in the first row of the table show the labor cost and travel cost per worker, total cost per worker, and total cost for three workers for one hour of labor.

    b. Row 5:

Hours	Labor Cost per Worker (in dollars)	Travel Cost per Worker (in dollars)	Cost per Worker (in dollars)	Total for Three Workers (in dollars)
1	37	25	62	186
2	74	25	99	297
3	111	25	136	408
4	148	25	173	519
5	185	25	210	630

15. a. Each row increases by $37, since this is the wage rate per hour.

    b. Answers will vary. Students may mention any of the following patterns:

Each row in column 1 increases by 1.

Each row in column 2 increases by $37.

Each row in column 3 is $25.

Each row in column 4 increases by $37.

Each row in column 5 increases by $111.

16. a. hours $\xrightarrow{\times \$37}$ ____ $\xrightarrow{+ \$25}$ ____ $\xrightarrow{\times 3}$

    b.

Hours	Labor Cost per Worker (in dollars)	Travel Cost per Worker (in dollars)	Cost per Worker (in dollars)	Total for Three Workers (in dollars)
1	37	25	62	186
2	74	25	99	297
3	111	25	136	408
4	148	25	173	519
5	185	25	210	630
6	222	25	247	741
7	259	25	284	852

17. No. Arrow strings cannot be used to make all tables. Only tables that have constant patterns can be represented by arrow strings. (See the next page for an example.)

Overview Students use arrow strings to describe patterns within a table.

About the Mathematics The table on the opposite page is also a ratio table. In contrast to the ratio table shown on page 33 of the Student Book, where each column represents a different number of servings, in this table, each row represents a different number of hours and the corresponding labor and travel costs.

Comments about the Problems

14. You may want to discuss how the table will change if, for example, the salary is increased to $40 per hour, the cost of travel is reduced to $18, or if there are only two workers involved.

15. Informal Assessment This problem assesses students' ability to interpret relationships displayed in tables.

16. Informal Assessment This problem assesses students' ability to interpret relationships displayed in tables, to use and interpret simple formulas, to generalize from patterns to symbolic relationships, and to use formulas in any representation (arrow language, arithmetic trees, words) to solve problems.

17. The purpose of this problem is to coax students to begin thinking about the different ways that tables are constructed. When the numbers in a table represent a pattern, that pattern can also be described with arrow language or a formula. The table on the next page does not contain a pattern and cannot be represented using string arrows or a formula.

Look carefully at the following table.

Typical Average Temperature (°F)	J	F	M	A	M	J	J	A	S	O	N	D
Acapulco	88	88	88	88	89	90	91	91	90	90	90	89
Antigua	80	80	80	82	90	90	90	90	89	89	89	83
Aruba	83	84	84	86	88	88	88	91	91	90	89	86
Cancún	84	85	88	91	94	92	92	91	90	88	86	82
Cozumel	84	85	88	91	94	92	92	91	90	88	86	82
Grand Cayman	88	87	86	88	88	89	90	91	91	89	88	88
Ixtapa	89	90	92	93	89	88	89	90	91	91	90	89
Jamaica	86	87	87	88	90	90	90	90	89	89	89	87
Los Cabos	73	74	79	83	88	93	95	93	92	89	82	74
Manzanillo	77	78	82	86	83	88	93	95	93	92	89	74
Mazatlán	73	74	79	83	84	92	94	92	92	90	85	71
Nassau	76	76	78	80	84	88	89	90	88	84	81	79
Puerto Vallarta	76	77	81	85	83	88	93	95	93	92	89	75
St. Martin/St. Kitts	80	81	82	83	86	86	86	87	87	86	85	84
U. S. Virgin Islands	80	81	82	83	88	88	90	90	88	87	86	86

18. a. Try to find the numbers and operations used to make this table. Explain what you found and if it makes sense.

b. Do you need to change your answer to problem **17?**

Summary

Some tables have regularities that can be described by a horizontal arrow string and a vertical arrow string. The strings can be useful in making or extending a table.

Be careful! Not every table has regularities.

Summary Questions

19. Find two tables from a newspaper or magazine, one table that has regularities and one that does not.

20. How can you tell whether arrow strings can be used to describe a table?

18. a. Answers and explanations will vary. Students' responses should indicate that since this table represents temperature data, there are no patterns or regularities in the table (since daily temperatures usually do not follow a pattern). The table cannot be represented using arrow strings.

b. Answers will vary. If students answer "yes" for problem **17,** this temperature table is an example that proves that their answer is incorrect.

19. Tables will vary.

Tables with collected data such as temperature, TV-ratings, and scores for sports events have no predetermined patterns.

Tables showing a ratio or representing a formula such as miles per gallon, price per unit, and the total for a bill have patterns.

20. Explanations will vary. Students' responses should include the following information stated in their own words:

Tables that contain a pattern and use the same arithmetic operation for both columns and rows can be described using arrow strings. Tables that do not contain a pattern cannot be described using arrow strings.

Overview Students analyze a table that is not generated by a formula and summarize the difference between this table and other tables in this section.

Planning You may want to use the table shown on the opposite page to discuss the main concept mentioned in the Summary: not all tables have regularity. Be sure to also include the following main points in your discussion:

- Arrow strings can be used to calculate ingredient amounts for different-size servings of a recipe.
- The numbers calculated in the arrow strings can be recorded in a table.
- If a table uses arrow strings that involve both addition and multiplication, there will be more than one solution for each table cell, depending on the order in which the calculations were done.
- Tables that have patterns and regularity can be represented using either horizontal or vertical arrow strings.

After students complete Section E, you may assign appropriate activities in the Try This! section, located on pages 54–58 of the Student Book, for homework.

Comments about the Problems

18–20. Some students may examine the numbers in the table before they notice the context of the table. The table contains temperatures for islands in the Caribbean and cities in Mexico that are considered vacation places because of the warm climate.

Writing Opportunity You may want students to use their answers to problem **20** as a journal entry.

SECTION F. ORDER OF OPERATIONS

Section Focus

Students use number sentences, arrow strings, and arithmetic trees to investigate the order of operations for a series of calculations. The instructional focus of Section F is to:

- Use order of operations to perform a series of calculations.
- Use arithmetic trees and parentheses to organize a series of calculations.

Planning Instruction

Day 15. Arithmetic Trees — Student pages 41–46

INTRODUCTION	Problems 1–3	■ Investigate calculations in which the operations of multiplication and addition are done in a different order.
CLASSWORK	Problems 4–10	■ Use order of operations and arithmetic trees to perform a series of calculations.
HOMEWORK	Problems 11–13	■ Represent plumbing repair jobs and stacking cups and chairs problems using arithmetic trees.

Additional Resources: Bringing Math Home, page 101; Extensions, pages 103 and 107; Writing Opportunity, page 105

Day 16. Flexible Computation — Student pages 47–49

INTRODUCTION	Problems 14 and 15	■ Compare and evaluate arithmetic trees.
CLASSWORK	Problems 16–20	■ Use arithmetic trees to make flexible calculations.
HOMEWORK	Problems 21–24	■ Calculate the cost of produce using arrow strings and arithmetic trees.

Day 17. Return to the Supermarket (Continued) — Student pages 50–53

INTRODUCTION	Problems 25 and 26	■ Create an arithmetic tree to calculate the cost of produce.
CLASSWORK	Problems 27–29	■ Use parentheses to replace arithmetic trees in calculations.
ASSESSMENT	Problems 30 and 31	■ Summary Questions.

Additional Resources: Extension, page 119; End-of-Unit Assessments, pages 124–133; Try This! Section F, Student page 58; *Balanced Assessment*

Materials

Student Resources	**Teacher Resources**	**Student Materials**
Quantities listed are per student.	No resources required.	Quantities listed are per student.
• Student Activity Sheet 5		• Calculator

*See Hints and Comments for optional materials.

Concept Development

Order of Operations

Section F introduces the order of operations to perform a series of calculations.

On **Day 15,** students investigate number sentences and arrow strings in which the operations of multiplication and addition are done in a different order. After concluding that these number sentences can be evaluated in different ways to get different answers, students learn the order of operations for addition, subtraction, multiplication, and division and apply this standard convention to a variety of calculations. As an introduction to the use of parentheses and as a way to represent calculations involving multiple operations, students investigate arithmetic trees to solve a string of calculations. Note: Arithmetic trees are introduced as a way to help students record the order of calculations for expressions that cannot be represented using a single arrow string.

On **Day 16,** students compare and evaluate arithmetic trees and use them as a tool for flexible calculations involving multiple operations. For homework, students calculate the cost of produce using arithmetic trees.

On **Day 17,** students create an arithmetic tree to calculate the cost of produce and use parentheses to replace arithmetic trees in calculations. Next, students demonstrate their ability to use the order of operations to perform a series of calculations.

Planning Assessment

Problems 16, 19, 24, 25, 31
• Describe and perform a series of calculations using an arithmetic tree.

Problems 16, 25, 31
• Use and interpret simple formulas.

Problem 16
• Rewrite numerical expressions to facilitate calculations.

Problems 19, 24, 25
• Use formulas to solve problems.

Problem 25
• Use word variables to describe a formula or procedure.

Problem 31
• Use conventional rules and grouping symbols to perform a sequence of calculations.
• Reason from a series of calculations to an informal formula.

F. ORDER OF OPERATIONS

Arithmetic Trees

While working on a problem from the previous section, a student named Enrique wrote the arrow string below. The problem was to find the cost of having three workers for two hours of repairs.

$$2 \xrightarrow{\times 37} \underline{\quad} \xrightarrow{+ 25} \underline{\quad} \xrightarrow{\times 3} \underline{\quad}$$

Karlene was working with Enrique, and she wrote:

$$2 \times 37 + 25 \times 3$$

She solved the problem and got an answer of 149. Her friend Enrique was very surprised.

1. a. How did Karlene get 149?

    b. Why was Enrique surprised?

Karlene and Enrique decided that the number sentence $2 \times 37 + 25 \times 3$ is not necessarily the same as the arrow string:

$$2 \xrightarrow{\times 37} \underline{\quad} \xrightarrow{+ 25} \underline{\quad} \xrightarrow{\times 3} \underline{\quad}$$

There is more than one way to interpret the number sentence. The calculations can be done in different orders.

2. Solve each of the problems below and compare your answers to those found by other students in your class.

    a. $1 + 11 \times 11$ **b.** $10 \times 10 + 1$

    c. $10 \xrightarrow{\times 10} \underline{\quad} \xrightarrow{+ 2} \underline{\quad}$ **d.** When can you be sure that everyone will get the same answer?

Sometimes the context of a problem helps you understand how to calculate it. For instance, in the repairs problem, Karlene and Enrique knew that the 3 represented the number of workers. So it makes sense to first calculate the subtotal of $2 \times 37 + 25$ and then multiply the result by three.

Sometimes people write their calculations for a problem in a very poor way:

$$2 \times 37 = 74 + 25 = 99 \times 3 = 297$$

3. Why is this a poor way to write the calculation?

1. a. Karlene probably multiplied 2 × 37 and 25 × 3 and added the two products to get an answer of 149.

b. Enrique may have expected Karlene to get the same answer that he did, 297. Enrique probably performed each operation from left to right, multiplying 2 × 37 to get 74, then adding 25 to get 99, and finally multiplying 99 by 3 to get an answer of 297.

Note: For the given context, 297 is the correct answer.

2. a. 122 or 132

b. 101 or 110

c. 102 is the only possible answer.

d. Everyone will get the same answer in calculations that use arrow language or if the order of operations is clearly defined.

3. This is a poor way to write the calculation because the equal signs may be misread. Someone could read the expression as 2 × 37 = 74 + 25 which is not true. Also, 74 + 25 does not equal 99 × 3 and so on.

Overview Students investigate calculations in which the operations of multiplication and addition are done in different orders to discover that the results are not the same. They revisit the misuse of the equal sign.

About the Mathematics In this section, the standard conventions regarding the order of operations are made explicit. In previous sections, students have intuitively discovered that the order of operations is not arbitrary, and this idea is formalized here. Only the four arithmetic operations of addition, subtraction, multiplication, and division are discussed. Other operations, such as powers and square roots, appear in the units *Powers of Ten* and *Building Formulas*.

As in reading, the logical order of performing operations in a string of calculations is to proceed from left to right. However, this linear sequence does not take into account that operations that appear later in the calculations should be done first. In long strings of calculations in which both addition and multiplication are involved, one needs to look at each operation in the whole string before doing any of the calculations.

Planning Students may work on problems **1–3** in small groups so that they see that the calculations will result in different answers, depending on the order in which the operations were performed.

Comments about the Problems

1. Doing the given calculation in sequence from left to right does not result in the correct answer since all multiplication operations should be done first. Let students struggle with this concept at this point.

2. If the multiplication operations appear first in the calculations (as in parts **b** and **c**), students' answers will be the same if they perform the calculations from left to right. When the multiplication operations appear later in the calculations, students' answers will differ.

3. As explained in Section A, the expression that appears on the left of the equal sign must be identical to the expression on the right of the equal sign for the math sentence to be true. This is not the case in the given expression used in this problem.

So that everyone can come up with the same answer to problems with different operations in them, mathematicians decided to treat multiplication and division as stronger than addition or subtraction. This means that you should do any multiplication or division before you add or subtract.

4. Use the mathematicians' way to find the value for each of the following expressions:

    a. $32 + 5 \times 20$

    b. $18 \div 3 + 2 \times 5$

    c. $47 - 11 + 6 \times 8$

Calculators and computers usually follow the mathematicians' rule. Some calculators, however, do not use the rule.

5. a. Use the mathematicians' rule to find $5 \times 5 + 6 \times 6$ and $6 \times 6 + 5 \times 5$.

    b. Does your calculator use the mathematicians' rule? How did you decide?

    c. Why do you think calculators have built-in rules?

It is important to have a way to write expressions so that it is clear which calculations to do first, second, and so on.

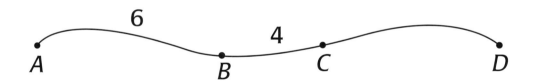

Above is a very simple map, not drawn to scale. Suppose someone told you that the distance from *A* to *D* is 15 miles. You can see in the drawing that the distance from *A* to *B* is 6 miles, and the distance from *B* to *C* is 4 miles.

6. What is the distance from *C* to *D*? Write down your calculations.

4. a. 132; 32 + 100 = 132.

b. 16; 6 + 10 = 16.

c. 84; 36 + 48 = 84.

5. a. $5 \times 5 + 6 \times 6 = 25 + 36 = 61$
$6 \times 6 + 5 \times 5 = 36 + 25 = 61$

b. Answers will vary, depending on the type of calculator students used.

c. Calculators have built-in rules so that users can get the same results.

6. 5 miles. Strategies will vary. Sample strategies:

Strategy 1

Some students may subtract: 15 − 6 − 4 = 5.

Strategy 2

Some students may first find a sum and then subtract: 6 + 4 = 10 and 15 − 10 = 5.

Strategy 3

Some students may use arrow language. Here are two possible ways:

$$15 \xrightarrow{-6} 9 \xrightarrow{-4} 5$$

$$6 \xrightarrow{+4} 10$$
$$15 \xrightarrow{-10} 5$$

Materials calculators (one per student)

Overview Students use the order of operations to determine if their calculators follow the same rules.

About the Mathematics When calculation strings involve both multiplication and division, the order in which these operations are performed is not important as long as multiplication and division operations are done before addition and subtraction operations. Calculation strings in which a combination of operations appear cannot be described using arrow strings. An arrow string can only be used to describe a linear sequence of calculations that proceed from left to right. On the next page, an alternative method to describe complex calculations is introduced.

Planning You may want to use the text on top of page 42 of the Student Book to discuss the rules for the order of operations.

Comments about the Problems

4. You may need to remind students to do the multiplication first for parts **a** and **c.** In part **b,** students can first calculate 18 ÷ 3, then 2 × 5, and add the results. It does not matter if multiplication or division is done first, as long as they are done before addition and subtraction.

5. Four-function calculators usually perform the calculations in the order in which the numbers are entered; all scientific calculators will use the order of operations rule.

6. Students may do the calculations in different stages, using the intermediate results in their next calculation.

Bringing Home the Math After students complete problem **5,** you might have them ask their parents or older brothers or sisters if anyone remembers the rules for the order of operations and if they were taught a sentence or phrase to help recall the rules. Many parents might remember a sentence, such as "Please excuse my dear Aunt Sally," in which the first letter of each word stands for one of the rules for the order of operations:

- Please: Parentheses first
- Excuse: Exponents second
- My: Multiplication and
- Dear: Division third
- Aunt: Addition and
- Sally: Subtraction last.

Although the above sentence suggests that multiplication should be done before division and that addition should be done before subtraction, the order in these two cases does not matter.

Telly found the distance from *C* to *D* by adding 6 and 4. Then she subtracted the result from 15. She could have used an *arithmetic tree* to write down this calculation.

• To make an arithmetic tree, begin by writing all the numbers.

15   6   4

• Then pick two numbers. Telly picked 6 and 4.

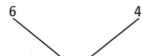

• Telly added the numbers and got 10.

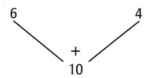

• Telly used the 15 and the new 10.

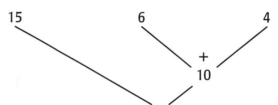

• She subtracted to get 5.

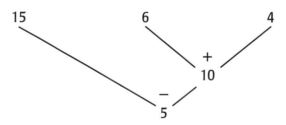

7. To help you understand arithmetic trees, complete those on **Student Activity Sheet 5.**

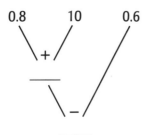 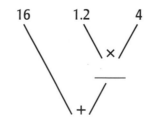 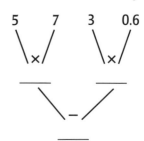

Solutions and Samples

of student work

7. a.

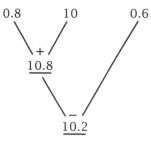

b.

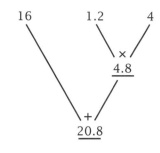

c.

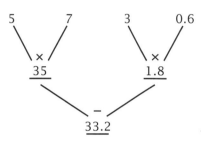

Materials Student Activity Sheet 5 (one per student); transparency, optional (one per class); overhead projector, optional (one per class)

Overview Students use arithmetic trees to describe and solve strings of calculations. This previews the need for parentheses.

About the Mathematics Arithmetic trees are introduced as a way to help students record the order of calculations for expressions that cannot be expressed using arrow strings. Expressions such as $2 \times 3 + 5 \times 8$ where 5×8 must be multiplied before the addition operation cannot be represented by a single arrow string. As students work through this section, they should realize some of the limitations of arrow strings and become more proficient in using arithmetic trees to solve calculation strings. Arithmetic trees are revisited in the unit *Building Formulas*.

Planning You might draw the given arithmetic trees on a transparency and have different students show how to work through each problem.

Comments about the Problems

7. Students are provided with a student activity sheet to solve arithmetic trees in this problem. However, for later problems in this section, students make their own arithmetic trees. Please make sure that students draw their trees neatly, especially when working with complicated trees.

Extension You may want to ask students to choose an arrow string from an earlier section of this unit and ask them to rewrite the string using an arithmetic tree.

8. Remember that multiplication is stronger than addition. Which of the following trees shows the proper calculation of $1 \times 2 + 3 \times 4$?

a.   **b.**   **c.**

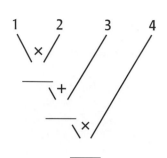   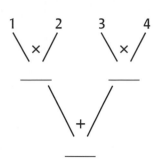   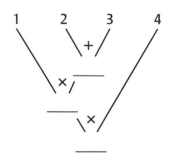

9. Which of the trees from problem **8** shows the same calculation as the arrow string below?

$$1 \xrightarrow{\times 2} \underline{\quad} \xrightarrow{+ 3} \underline{\quad} \xrightarrow{\times 4} \underline{\quad}$$

10. At the beginning of this section, two students—Karlene and Enrique—got different answers for the cost of a two-hour job.

    a. Does the tree below give the same answer that Karlene got?

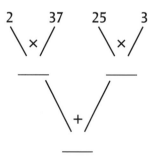

    b. Draw a tree that gives the correct answer for a two-hour job.

8. Arithmetic tree **b** shows the proper calculation. All multiplication calculations are done first, followed by the addition calculation.

9. Arithmetic tree **a.**

10. a. Yes; this tree gives the same result that Karlene got (149).

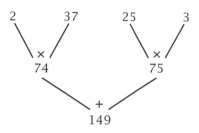

b.

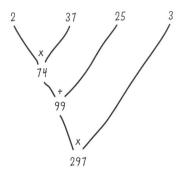

Overview Students solve more calculation strings using arithmetic trees and compare them to arrow strings.

About the Mathematics An arithmetic tree consists of "branches." Each branch connects two numbers with an operation. The level of the connection of the branches indicates the order of the operations. A higher connection of two branches indicates an operation to be done before a lower connection.

Comments about the Problems

8. If students do not know how to approach this problem, you may want to ask, *How do you know which operation is to be performed first in an arithmetic tree?* [The ones on a higher level are performed before the ones below them.] In this arithmetic tree, the addition operation is to be done first, followed by the two multiplication calculations. Although the tree suggests that the multiplication calculation on the left be done before that shown on the right, these two multiplications can be done in any order.

9. As explained in the About the Mathematics section above, the structure of the arithmetic tree in problem **8a** indicates that the operation at the far left should be done first since it has the highest position in the tree, followed by the operations in the middle and the far right of the arithmetic tree.

Writing Opportunity Ask students to write two paragraphs about the advantages and disadvantages of arithmetic trees and arrow strings in their journals. You may collect students' work to assess their understanding of arithmetic trees.

One student feels that the calculation for repair bills (shown on pages 38 and 39) should start with the travel costs since the customer always has to pay them. Another student thinks that it is impossible to start an arrow string with the travel costs.

11. Is it impossible to start an arrow string with the travel costs? Explain.

An arithmetic tree can have the travel costs first.

12. Below is an arithmetic tree for calculating the bill for a two-hour job with the travel costs first. Does this arithmetic tree give the correct total (as shown in the table on page 39)?

11. Yes, it is impossible to start an arrow string with the travel costs. Explanations will vary. Some students may reason that if the travel cost is moved to the beginning of the arrow string, the addition calculation must be performed before the multiplication calculation, which means that the order of operations is not being followed. Others may reason that arrow strings can only be done in one order (from left to right). If the travel cost is moved to the start of the string, it will result in an incorrect answer.

12. Yes. This arithmetic tree gives the correct total. It follows the same order as the original arrow string.

Overview Students revisit the arrow string that describes the repair job bills and investigate whether or not a given arithmetic tree gives the correct answer for this situation.

Comments about the Problems

11. The multiplication (number of hours times $37) should be done before the $25 is added. Arrow strings can only be done in order from left to right. If the travel cost is moved to the beginning of the arrow string, it will result in an incorrect answer.

12. This arithmetic tree clearly shows that the multiplication operation (2 times $37) is to be done first.

Extension Students may look in magazines or newspapers for formulas. You may ask students to try to describe them using arrow language or arithmetic trees.

It is possible to build an arithmetic tree that uses words instead of numbers. In the left column of the table below, a word tree has been built to describe the repair bill. In the right column, numbers have been used.

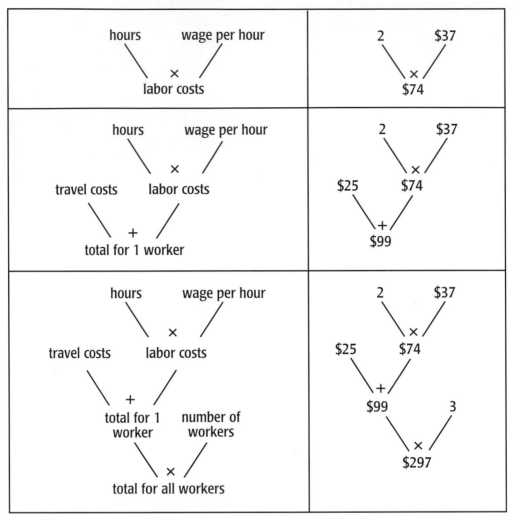

By making some of the "branches" longer, the tree looks like the one from problem **12.**

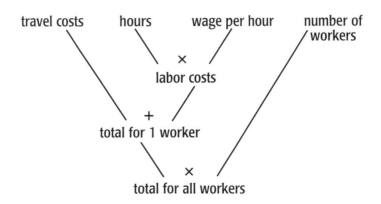

13. Make a word tree that shows how to find the height of a stack of cups or chairs.

13. Arithmetic tree for a stack of chairs:

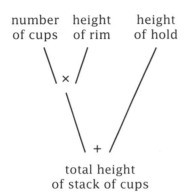

number of chairs   1 (first chair)   height added by each additional chair   height of 1 chair

×

+

total height
of stack of chairs

Arithmetic tree for a stack of cups:

number of cups   height of rim   height of hold

×

+

total height
of stack of cups

Materials transparency, optional (one per class); overhead projector, optional (one per class)

Overview Students analyze a word tree that describes the formula for the repair bill from Student Book pages 38 and 39. Then they use an arithmetic tree of words to describe the general formula for the height of a stack of chairs or cups.

About the Mathematics An arithmetic tree that uses words instead of numbers is another way to represent a formula. The words in the tree describe the numbers that will replace them. These words may be referred to as *variables*. The variables on top of the tree are the input variables, and the variable at the bottom of the tree is the output variable. By assigning a value to each input variable, one can perform the calculations in the tree to get the output value.

Have students use words or terms that make sense to them to describe the relationships in a word tree. The idea of describing a relationship in general terms using words (instead of numbers) may be too abstract for some students of this age.

Planning You may want to reproduce the word tree for the formula of the repair bill onto a transparency to facilitate a discussion of the meaning of each part of the tree. Have students tell in their own words what each place in the tree stands for.

Comments about the Problems

13. A word tree is used here to show the relationship between the formula and the arithmetic operations involved and to help students see how words (or variables) can be used to generalize the sequence of computations.

Flexible Computation

Arithmetic trees can be used to help make some addition and subtraction problems easier.

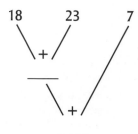 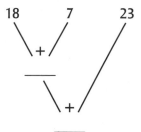

14. a. Compare the two trees above.

   b. Draw a tree that makes adding the three numbers in part **a** easier.

Addition problems with more numbers have many possible arithmetic trees. Below are two trees for $\frac{1}{2} + \frac{1}{4} + \frac{3}{4} + \frac{3}{2}$.

15. a. Copy the trees and find the sum.

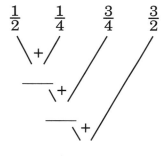 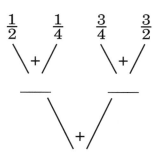

   b. Design two other arithmetic trees for the same problem and solve them.

   c. Which arithmetic tree makes $\frac{1}{2} + \frac{1}{4} + \frac{3}{4} + \frac{3}{2}$ easiest to compute? Why?

14. a. Answers will vary. Both trees give the same output of 48. The numbers and operations are the same. The order in which the numbers are added differ. In each tree, 18 is added first.

b.

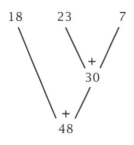

15. a. The sum is 3.

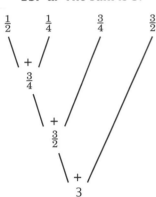 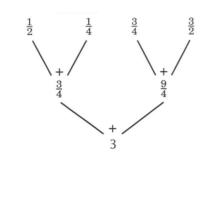

b. Answers will vary. Sample trees:

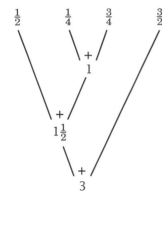 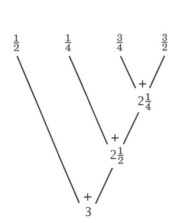

c. Answers will vary, depending on the trees that students chose to draw. Most likely, the trees in which $\frac{1}{4}$ and $\frac{3}{4}$ were added, $\frac{1}{2}$ and $\frac{3}{2}$ were added, and the two results totaled would make the fractions easiest to compute.

Overview Students experiment with reorganizing arithmetic trees to find ways to make the computations easier to solve.

About the Mathematics Since the operations on this page are restricted to addition, the arithmetic trees can be reorganized in any order. Students must use their number sense to evaluate a preference for one arithmetic tree over another.

Comments about the Problems

14. Encourage students to group numbers in ways so that the numbers can be added mentally. Adding 23 + 7 results in a number whose ones digit is 0, which facilitates adding it to any other number mentally.

15. Combining fractions by using pairs that combine easily may be new to students. Some students may need to review some of the common fraction sums. Fractions are studied extensively in the units *Some of the Parts* and *Fraction Times*. For additional resources, refer to *Number Tools* Volumes 1 and 2.

16. Design an arithmetic tree that will make each of the following problems easy to solve.

    a. $7 + 3 + 6 + 4$

    b. $4.5 + 8.9 + 5.5 + 1.1$

    c. $\frac{4}{10} + \frac{1}{2} + \frac{1}{10} + \frac{3}{4}$

17. How are different arithmetic trees for the same problem the same? How can they differ?

You may have noticed that if a problem has only addition, the answer is the same no matter how you make the arithmetic tree. You may wonder if this is true for subtraction.

18. Do the following trees give the same result?

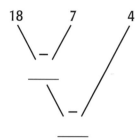 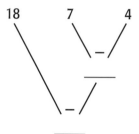

Near the beginning of the section, Telly used the following tree to find the distance from point *C* to point *D*:

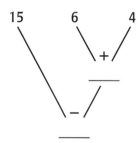

19. Make another arithmetic tree with 15, 6, and 4 across the top that gives the same result.

20. Calculate $176 - 89 - 11$ and describe what you did. (*Hint:* If you look at the previous problem, you may think of an easy way to calculate this one.)

16. Answers for parts **a–c** will vary. Sample responses:

a.

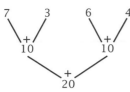

b.

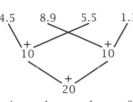

c.

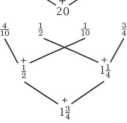

17. Answers will vary. Students' responses should indicate that the trees for the same problem are the same in that they give the same result. They are different in the order in which the calculations are performed.

18. No, the trees do not give the same result. The first one gives an answer of 7, and the second one gives an answer of 15.

19. Answers will vary. Sample response:

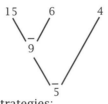

20. 76. Strategies will vary. Sample strategies:

Strategy 1

Some students may subtract the sum of the two numbers.

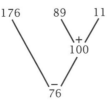

Strategy 2

Some students may subtract each number individually.

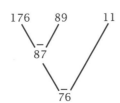

Overview Students work on more problems involving arithmetic trees.

Comments about the Problems

16. Informal Assessment This problem assesses students' ability to describe and perform a series of calculations using an arithmetic tree, to use and interpret simple formulas, and to rewrite numerical expressions to facilitate calculation. Students should be able to justify why they think their arithmetic trees make the problems easier to solve.

19. Informal Assessment This problem assesses students' ability to describe and perform a series of calculations using an arithmetic tree and to use and interpret formulas to solve problems.

Return to the Supermarket

TOMATOES $1.50/lb

GRAPES $1.70/lb

GREEN BEANS $0.90/lb

SORRY— OUT OF ORDER SEE CASHIER

The machine at Veggies-R-Us is broken. Ms. Prince buys 0.5 pound of grapes and 2 pounds of tomatoes.

21. How much does she have to pay?

22. Can you write an arrow string to show how to figure Ms. Prince's bill? Why or why not?

23. Can Ms. Prince's bill be calculated with an arithmetic tree? If so, make the tree. If not, explain why not.

Dr. Keppler buys 2 pounds of tomatoes, 0.5 pound of grapes, and $\frac{1}{2}$ pound of green beans.

24. Make an arithmetic tree for the total bill.

21. $3.85. Strategies will vary. Sample strategy:

Strategy 1

Some students may calculate the tomatoes and grapes individually and then add the products.

Grapes:	$0.5 \times \$1.70 = \$0.85$
Tomatoes:	$2 \times \$1.50 = \$3.00$
Total:	$\$0.85 + \$3.00 = \$3.85$

22. No. Explanations will vary. Some students may reason that there is no way to write one arrow string for an expression in which two pairs of numbers are multiplied and then their products are added.

23. Yes. Ms. Prince's bill can be calculated using an arithmetic tree.

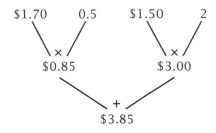

24.

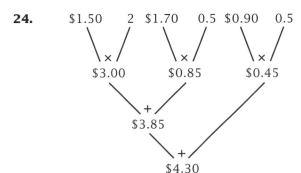

Overview Students revisit the context of Veggies-R-Us from Section C and investigate how arrow strings and arithmetic trees can be used to describe the calculations.

About the Mathematics In Section C, an arrow string was used for each single calculation. In this section, the calculations are combined by using an arithmetic tree.

Comments about the Problems

24. Informal Assessment This problem assesses students' ability to describe and perform a series of calculations using an arithmetic tree and to use formulas to solve problems.

Note that the addition operations can be done in any order, as long as the multiplication operations are done first.

25. Make a tree that can be used for any combination of tomatoes, grapes, and green beans. Use the words "weight of tomatoes," "weight of grapes," and so on.

The store manager gave each of the cashiers a calculator that uses the rule that multiplication is stronger than addition. Then she wrote these directions.

amount of tomatoes × 1.50 + amount of grapes × 1.70 + amount of green beans × 0.90 =
    (in pounds)   (in pounds)   (in pounds)

26. If the cashiers punch in a calculation using the directions above, will they get the correct total for the bill?

25. Answers will vary. Sample response:

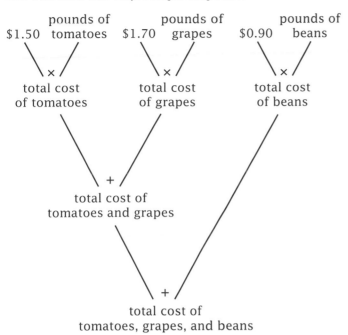

26. Yes, this will give a correct total because the calculator multiplies before adding.

Overview Students make an arithmetic tree using words that could be used to calculate the cost of produce purchased at Veggies-R-Us.

Comments about the Problems

25. Informal Assessment This problem assesses students' ability to describe and perform a series of calculations using an arithmetic tree, to use and interpret simple formulas, to use word variables to describe a formula or procedure, and to use formulas to solve a problem.

Students should understand how to use the order of operations here and how the results might differ from calculations performed on the same number using arrow strings.

26. The string of calculations will give the correct amount when key entered into any calculator that uses the rule of the order of operations. However, pushing ENTER after entering each number produces the same result as with arrow strings, and the result will not be correct.

WHAT COMES FIRST?

Arithmetic trees are useful because there is no question about the order of the calculation. The problem is that they take up a lot of room on paper. Copy the first tree on the right.

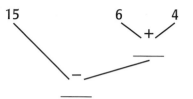

Since the 6 + 4 is done first, circle it on your copy.

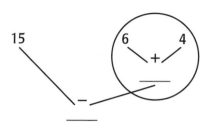

The tree can then be simplified to:

Instead of the second arithmetic tree, you could write: 15 − ⬭ 6 + 4

27. What does the circle mean?

The whole circle is not necessary. People often write 15 − (6 + 4). This does not take up much space, but it shows the order with the parentheses.

28. a. Rewrite the tree on the right using parentheses.

   b. Make a tree for 3 × (6 + 4).

   c. What is the value of 5 × (84 − 79)?

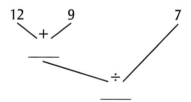

   d. Rewrite the tree on the lower right using parentheses.

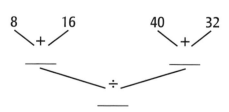

29. Rewrite Karlene's problem from page 41: 2 × 37 + 25 × 3. Use parentheses so that the correct total for the bill will be calculated.

27. The circle indicates that 6 + 4 is computed first.

28. a. $(12 + 9) \div 7$

b.

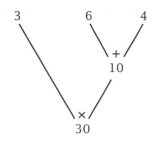

c. 25

d. $(8 + 16) \div (40 + 32)$

29. $(2 \times 37 + 25) \times 3$ or
$((2 \times 37) + 25) \times 3$

Overview Students make the transition from using arithmetic trees to using parentheses.

About the Mathematics Arithmetic trees visually show students how to perform the calculations needed to solve a problem. However, arithmetic trees take up a lot of space. Instead of drawing "elaborate" trees, students can circle the calculation that is to be done first. The advantage of circles is that the whole series of calculations can be written in one line. Students then make the transition from circles to parentheses. If students have difficulty using the more abstract parentheses, allow them to use arithmetic trees or circles.

Planning Be sure that students understand the role of parentheses, and that parentheses can be used to indicate the order of operations the same way as arithmetic trees.

Comments about the Problems

28. Students should recognize that the operation within the parentheses should be done first, and that this operation corresponds to a high position in the arithmetic tree.

29. As explained in the solutions column, students may use parentheses in either of two ways to show the order of operations.

Extension Have students look back at earlier problems in this unit and rewrite the arrow strings and number expressions using parentheses.

Summary

The beginning of this unit focused on formulas using arrow language.
There are also other ways to write formulas.

You can write them with words:

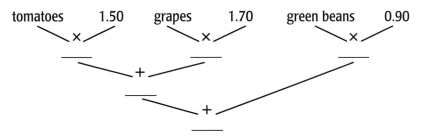

cost = tomatoes × $1.50 + grapes × $1.70 + green beans × $0.90
      (in lb)   (in lb)   (in lb)

You can use arithmetic trees:

tomatoes   1.50   grapes   1.70   green beans   0.90

Arithmetic trees show the order of calculation. If a problem is not in an
arithmetic tree and does not have parentheses, there is a rule for the
order of operations: Do the multiplication and division first.

1 × 2 + 3 ÷ 4 − 5 is calculated as follows:

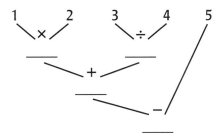

Parentheses can be used to convert from an arithmetic tree to an
expression that shows which operations to do first.

About the Mathematics In this section, students have worked with different ways to describe formulas:

- in words written in one line with an equal sign and operation signs (using parentheses, if necessary).

- as an arithmetic tree, positioning the calculations to be done first high in the tree.

The arithmetic tree can be used as a visual representation of any formula.

Students have also learned about the standard order of operations and how to use parentheses to group calculations that need to be done first in a calculation string.

Planning You may review the main math concepts covered in this section by discussing the text in the Summary. There are no problems to solve on this page.

Summary Questions

30. Write a story in your journal in which the calculations below are used.

    a. $(15 \times 16 + 20) \times 5$

    b. $15 \times 16 + 20 \times 5$

31. Phoebe wants to make an arithmetic tree using the numbers 421, 17, 45, and 23. She knows that the calculation $421 - 40 - 45$ is an intermediate step.

    a. Draw Phoebe's arithmetic tree.

    b. Rewrite the arithmetic tree you drew as an expression with parentheses.

30. a. Stories will vary. Sample story:

Five friends each purchased 16 apples at 15¢ apiece and a pencil for 20¢. How much did they spend altogether? [Solution: (15 × 16 + 20) × 5 = (240 + 20) × 5 = 260 × 5 = 1300¢ or $13.00]

b. Stories will vary. Sample story:

Sarah purchased 16 apples at 15¢ apiece and 5 pencils at 20¢ each. How much did she spend altogether? [Solution: 15 × 16 + 20 × 5 = 240 + 100 = 340¢ or $3.40]

31. a. Answers will vary.

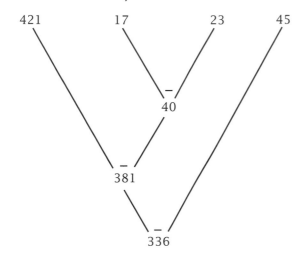

b. 421 − (17 + 23) − 45 = 336 or
(421 − (17 + 23)) − 45 = 336 or
421 − ((17 + 23) + 45) = 336

Overview Students write a story in their journal that corresponds with a given calculation string. They also draw arithmetic trees and rewrite trees as expressions with parentheses.

Planning After students complete Section F, you may assign appropriate activities in the Try This! section, located on pages 54–58 of the Student Book, for homework.

Comments about the Problems

31. Informal Assessment This problem assesses students' ability to describe and perform a series of calculations using an arithmetic tree; to use and interpret simple formulas; to use conventional rules and grouping symbols to perform a sequence of calculations; and to reason from a series of calculations to an informal formula.

Assessment Overview

Students work on three assessment activities that can be used at the end of the unit. You can evaluate these assessments to determine what each student knows about the main concepts in the unit.

Goals

- describe and perform a series of calculations using an arrow string

- describe and perform a series of calculations using an arithmetic tree

- use and interpret simple formulas

- use conventional rules and grouping symbols to perform a sequence of calculations

- use reverse operations to find the input for a given output

- rewrite numerical expressions to facilitate calculation

- reason from a series of calculations to an informal formula

- interpret relationships displayed in tables

- use word variables to describe a formula or procedure

- generalize from patterns to symbolic relationships

- solve problems using the relationship between a mathematical procedure and its inverse

- use formulas in any representation (arrow language, arithmetic tree, words) to solve problems

Assessment Opportunities

Get Your Programs Here!, p. 142
Forest Fire Fighting, pp. 143 and 144
Expressions, p. 145

Get Your Programs Here!, p. 142
Expressions, p. 145

Get Your Programs Here!, p. 142
Forest Fire Fighting, pp. 143 and 144
Expressions, p. 145

Expressions, p. 145

Get Your Programs Here!, p. 142
Forest Fire Fighting, pp. 143 and 144

Expressions, p. 145

Get Your Programs Here!, p. 142
Expressions, p. 145

Get Your Programs Here!, p. 142
Expressions, p. 145

Forest Fire Fighting, pp. 143 and 144

Get Your Programs Here!, p. 142
Forest Fire Fighting, pp. 143 and 144
Expressions, p. 145

Get Your Programs Here!, p. 142
Forest Fire Fighting, pp. 143 and 144

Get Your Programs Here!, p. 142
Forest Fire Fighting, pp. 143 and 144

Pacing

When combined, the three assessment activities will take approximately one 45-minute class session. For more information on how to use the three problems, see the Planning Assessment section on the next page.

About the Mathematics

These three end-of-unit activities assess the majority of the goals of the *Expressions and Formulas* unit. Refer to the Goals and Assessment Opportunities section on the previous page for information regarding the specific goals that are assessed in each activity. Some of the tasks assess more than the mathematical content of the unit. In the Forest Fire Fighting assessment, students must demonstrate understanding of the appropriateness of formulas used to describe a situation.

Materials

- Assessments, pages 142–145 of this Teacher Guide (one of each per student)
- calculators, pages 127, 129, 131, and 133 of the Teacher Guide, optional (one per student)

Planning Assessment

These three activities can be used as one end-of-unit assessment.

The activities are designed for individual assessment but some problems can be done in pairs or in small groups. It is important that the students work individually if you want to evaluate each student's understanding and abilities.

Make sure that you allow enough time for students to do the assessments. If the students need more than one class session, it is suggested that they finish during the next mathematics class or you may assign an activity as homework. The students should be free to solve the assessment problems in their own way. Calculators may be used if the students choose to use them.

If individual students have difficulties with any particular problem and solve it incorrectly, you may give the student the option of making a second attempt after giving him or her a hint. Suggestions for these hints are described in the Hints and Comments columns. You may also decide to use one of the optional problems or Extension activities not previously done in class as additional assessments for students who need additional help.

Scoring

Answers are provided for all assessment problems. The method of scoring the problems depends on the types of questions in each assessment. Most questions require students to explain their reasoning or justify their answers. For these questions, the reasoning used by the students in solving the problems as well as the correctness of the answers should be considered as part of your grading scheme. A holistic scoring scheme can be used to evaluate an entire task. For example, after reviewing a student's work, you may assign a key word such as *emerging, developing, accomplishing,* or *exceeding* to describe their mathematical problem-solving, reasoning, and communication.

On other tasks, it may be more appropriate to assign point values for each response. Student progress toward goals of the unit should also be considered. Descriptive statements that include details of a student's solution to an assessment activity can be recorded. These statements would provide insight into a student's progress toward a specific goal of the unit. Descriptive statements can be more informative than recording only a score and can be used to document student growth in mathematics over time.

GET YOUR PROGRAMS HERE!

Use additional paper as needed.

Jim is selling programs at a Cubs' game.
He is paid $5.00 per game. In addition,
he gets $0.10 for each program he sells.

1. a. What are his total earnings
if he sells 75 programs at one game?

   b. What are his total earnings if he sells
   200 programs at one game?

2. If you know the number of programs Jim sold
one day, write a formula to calculate his total earnings for that game.

3. At the first game of the season he made $18.60. How many programs did he sell?

4. Jim's friend sold 350 programs. Make an arithmetic tree using the numbers
shown below to find how much his friend earned.

$5.00   350 $0.10

Solutions and Samples
of student work

Hints and Comments

Get Your Programs Here!

1. **a.** $12.50. Strategies will vary. Sample strategy:

   $75 \times \$0.10 + \$5.00 = \$12.50$

   b. $25. Strategies will vary. Sample strategy:

   $\$200 \times \$0.10 + \$5.00 = \$25.00$

2. Answers will vary. Sample responses:

   Using an arrow string:

   number of programs $\xrightarrow{\times\, 0.10}$ _____ $\xrightarrow{+\, 5}$ total earned

   Using an arithmetic tree:

   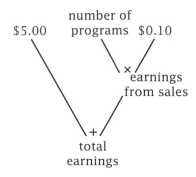

   Using a word formula:

   $0.10 \times$ number of programs $+ 5 =$ total earnings

3. 136 programs

4.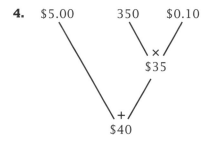

   Jim's friend earned $40.

Materials Get Your Programs Here! assessment, page 142 of the Teacher Guide (one per student); calculators, optional (one per student)

Overview Students write a formula, make an arithmetic tree, and do some calculations to determine the total earnings from selling programs at a baseball game.

About the Mathematics In this activity, students show their understanding and use of the order of operations, arrow language, arithmetic trees, and formulas to describe a situation.

Planning This activity consists of four questions and can be solved in about 15 minutes. You might want students to work individually on these problems.

Comments about the Problems

1. These two questions ask students to do some calculations using any strategy they prefer. Be sure that students show how they got their answers.

2. In this problem, students express a calculation rule as a formula. They can choose arrow strings, arithmetic trees, or word formulas.

3. In this problem, students must find the input for a given output. Some students may solve this problem using reverse arrow strings. Others may simply calculate to find the answer.

Extension You may ask students to think of a similar situation (selling items) and have them describe that situation using arrow strings, arithmetic trees, and word formulas.

Use additional paper as needed.

To be able to fight fires in a large National Park, several airfields are built. From each airfield, airplanes carrying loads of water can take off to fight a fire. It is important that the planes reach the fire within a reasonable amount of time.

There is a formula that describes the average time it takes to get to a fire:

$t = d \div s + h$

t = total time
d = distance in miles
s = speed in miles per hour
h = handling time in hours (the time between a call for help and the takeoff)

1. a. Write the formula in arrow language.

   b. Explain the formula.

2. Use the back of this page to calculate the total time to get to a fire if:
   • it is 100 miles from the airfield;
   • the plane flies at an average speed of 50 miles per hour;
   • the handling time is 0.25 hour.

1. a. $d \xrightarrow{\div s} \underline{\qquad} \xrightarrow{+ h} t$

b. Explanations will vary. Students' explanations should include information about how the distance is divided by the speed and that result is added to the handling time to get the total time needed to get to a fire.

2. 2.25 hours. Strategies will vary. Sample strategy:

$t = d \div 5 + h$
$ = 100 \div 50 + 0.25$
$ = 2 + 0.25$
$ = 2.25$

Materials Forest Fire Fighting assessment, page 143 of the Teacher Guide (one per student); calculators, optional (one per student)

Overview Students explore, explain, and use a formula that describes the average time it takes an airplane to get to a fire.

About the Mathematics In these assessment problems, students must demonstrate that they have developed a critical attitude toward formulas. Students are asked to determine how useful a formula might be and if the formula always works. Students also describe and explain the given formula using an arrow string and perform calculations to compute the output, as well as the reverse arrow string calculations to compute the input, given the output.

Planning The first two problems help students to become familiar with the context and the basic formula used within this context. Problems **3** and **4** on the next page are of a more complex nature. You might want students to work individually on these problems.

Comments about the Problems

1. This problem assesses students' ability to describe a formula using arrow strings.

2. Students may use the arrow string from problem **1** or the given formula to solve this problem. Others may use their own strategy.

Use additional paper as needed.

3. What is the maximum distance you can have from any point in the park to an airfield if you want the maximum total time needed to arrive at the scene to be 3 hours and 45 minutes? The handling time is always 0.25 hour and the average speed of the plane is 50 mph.

Last week a plane had to go to a fire only 150 miles from the airfield. However, it took 10 hours to get there.

4. a. Does this mean that the formula has to be changed?

b. Describe the factors that can influence the time needed and how they can be avoided.

c. If the total time has to be shortened, where could one save time, the handling time or the average speed?

3. 175 miles

4. a. No. The formula does not necessarily have to be changed. The long time on the given day could be due to a weather condition or a long handling time.

   b. Answers will vary. Sample responses:

   • slow speed—could be improved with new planes or better maintenance

   • long handling time—could be improved by increasing the size of the ground crew, providing more training for the crew, or having frequent drills

   c. Answers will vary. Handling time is only 15 minutes, so increasing the average speed of the plane would help save time.

Materials Forest Fire Fighting assessment, page 144 of the Teacher Guide (one per student); calculator, optional (one per student)

Overview Students continue working with the formula that describes the average time it takes to get to a fire. These problems extend the work done in this unit. The formula is given using variables.

Planning You may want students to work individually on these problems.

Comments about the Problems

3. As in problem **2**, the output (time) is given, and students are to find the input (distance). Several strategies can be used: ratio tables, the guess and check method, logical reasoning, or plugging the given values into the formula and working backwards.

4. Students need to consider the appropriateness of the given formula here. If students are having difficulty, ask: *Can the given formula be used to describe the situation in this problem? If not, why?* [See the answer in the solutions column.]

Writing Opportunity You may ask students to write two paragraphs in their journals describing the pros and cons of using the given formula in problem **4.**

EXPRESSIONS

Use additional paper as needed.

Compare the following mathematical expressions:

a. $4 \xrightarrow{\times 5} \ldots \xrightarrow{+ 5} \ldots \xrightarrow{\times 4} \ldots$

b.
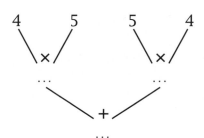

c. $4 \times 5 + 5 \times 4 = \ldots$
(as the calculator you used earlier would compute it)

1. Do these expressions give the same or different results? Explain why.

2. Rewrite part **c** as an expression using parentheses.

Solutions and Samples
of student work

1. Expressions **b** and **c** give the same result because the operations take place in the same order. Expression **a** gives a different result because the addition calculation occurs between the two multiplication calculations.

    a. 100

    b. 40

    c. 40

2. $(4 \times 5) + (5 \times 4)$

Hints and Comments

Materials Expressions assessment, page 145 of the Teacher Guide (one per student); calculators, optional (one per student)

Overview Students determine if three numerical expressions (described using an arrow string, arithmetic tree, and a calculation string) give the same results.

About the Mathematics This assessment evaluates students' understanding of the order of operations and of the different representations for writing and performing calculations (arrow language, arithmetic trees, and algebraic expressions).

Comments about the Problems

1. Students must compare the different representations and show an understanding of the order of operations.

    They may use examples with numbers, or reason on a more general level about the representations. Students who use a general approach, such as word formulas, should receive more credit than those who use examples with numbers.

2. This problem assesses students' understanding of the order of operations and use of parentheses.

Expressions and Formulas
Glossary

The Glossary defines all vocabulary words listed on the Section Opener pages. It includes the mathematical terms that may be new to students, as well as words having to do with the contexts introduced in the unit. (Note: The Student Book has no glossary in order to allow students to construct their own definitions, based on their personal experiences with the unit activities.)

The definitions below are specific to the use of the terms in this unit. The page numbers given are from this Teacher Guide.

arithmetic tree (p. 102) a mathematical map showing the order for performing calculations

arrow language (p. 8) a method of writing a calculation by showing each operation with an arrow

arrow string (p. 12) a calculation that is written using arrow language where more than one arrow is used

conversion rule (p. 68) a rule that describes how a quantity measured in one magnitude can be changed to another magnitude (e.g., exchanging money, weight, length, volume)

entries (p. 92) the numbers in a table or chart

formula (p. 46) directions in the form of arrow strings, arithmetic trees, or words that tell how to do a specific type of calculation which can be used repeatedly for different numbers; a formula describes the relation between variables

hold (p. 48) the distance from the bottom of the cup to the bottom of the rim

input (p. 40) a starting number for a series of calculations or for a formula

opposite operation (p. 76) multiplication and division are opposite operations and addition and subtraction are opposite operations; the opposite operation is also known as the inverse operation

order of operations (p. 98) a set of rules that predetermines the order in which a sequence of calculations is to be performed: all multiplication and division is done before addition and subtraction, always working from left to right

output (p. 40) the resulting number, or answer, for a series of calculations or for a formula

parentheses (p. 118) mathematical symbols used to show the order in which the operations are to be performed; grouping symbols

reverse string (p. 72) an arrow string that works through a calculation backwards, and is read from right to left

small-coins-and-bills-first method (p. 24) a strategy for giving change

thickness (p. 74) the diameter of an object

timetable (p. 82) a table or chart that shows arrival and departure times

word formula (p. 109) a formula written in words

Blackline
Masters

Dear Family,

Your child is about to begin the *Mathematics in Context* unit called *Expressions and Formulas*.

Students use "arrow language" to represent a sequence of arithmetic operations. Arrow language allows students to write out a long calculation and keep track of intermediate results. After completing Section B, your child can demonstrate arrow language to show how much change you should expect when making a purchase.

After becoming comfortable with arrow language, students are introduced to basic formulas. They calculate prices at a grocery store and the cost of a taxi ride using formulas. After completing Section C, your child can explain how to use a formula to figure out what size bike he or she needs.

Students also identify patterns in tables and then use the patterns to develop formulas that allow them to extend the tables. You may want to study a bus or train schedule with your child to look for patterns in the timetables.

You might also discuss formulas that are used in everyday situations, such as a calculation rule you may use to determine the amount of paint needed for a room or a formula you use in your line of work.

We hope you enjoy working with your child in *Expressions and Formulas*.

Sincerely,

The Mathematics in Context Development Team

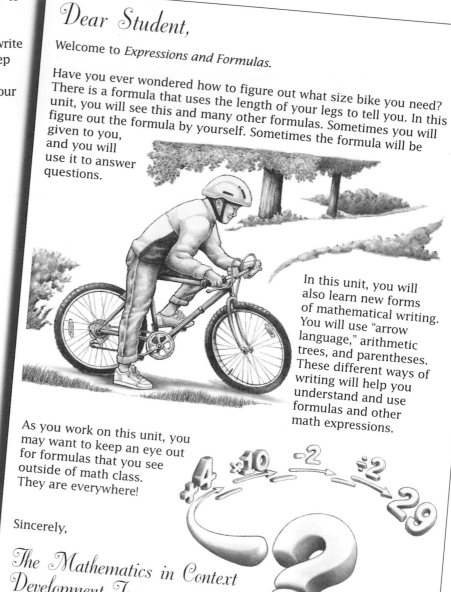

Dear Student,

Welcome to *Expressions and Formulas*.

Have you ever wondered how to figure out what size bike you need? There is a formula that uses the length of your legs to tell you. In this unit, you will see this and many other formulas. Sometimes you will figure out the formula by yourself. Sometimes the formula will be given to you, and you will use it to answer questions.

In this unit, you will also learn new forms of mathematical writing. You will use "arrow language," arithmetic trees, and parentheses. These different ways of writing will help you understand and use formulas and other math expressions.

As you work on this unit, you may want to keep an eye out for formulas that you see outside of math class. They are everywhere!

Sincerely,

The Mathematics in Context Development Team

Ingredients	4 servings	2 servings	6 servings
Flour	250 g		
Butter	125 g		
Eggs	1		
Apples	450 g		

Ingredients	Servings							
	2	3	4	5	6	8	10	12
Flour			250 g					
Butter			125 g					
Eggs			1					
Apples			450 g					

Use with *Expressions and Formulas,* pages 34 and 35.

O'HARE TO NORTHWEST INDIANA

Reading schedules from O'Hare: Find the time you leave O'Hare in the left column. Read straight across to the right on the same line to your destination point. This will show your arrival time at that point.

Leave O'Hare Lower Terminal	Arrive Expo Center	Arrive Hammond/ Highland	Arrive Glen Park	Arrive Merrillville
5:45 A.M.		7:05 A.M.	7:20 A.M.	7:40 A.M.
6:45 A.M.		8:05 A.M.	8:20 A.M.	8:40 A.M.
7:45 A.M.		9:05 A.M.	9:20 A.M.	9:40 A.M.
8:45 A.M.		10:05 A.M.	10:20 A.M.	10:40 A.M.
9:45 A.M.		11:05 A.M.	11:20 A.M.	11:40 A.M.
10:45 A.M.		12:05 P.M.	12:20 P.M.	12:40 P.M.
11:45 A.M.		1:05 P.M.	1:20 P.M.	1:40 P.M.
12:45 P.M.		2:05 P.M.	2:20 P.M.	2:40 P.M.
1:45 P.M.		3:05 P.M.	3:20 P.M.	3:40 P.M.
2:45 P.M.		4:05 P.M.	4:20 P.M.	4:40 P.M.

7.

```
        +2    +2    +2
   ┌─────┬─────┬─────┬─────┐
   │  5  │     │     │     │
+3 │     │     │     │     │
   ├─────┼─────┼─────┼─────┤
+3 │     │     │     │     │
   ├─────┼─────┼─────┼─────┤
+3 │     │     │     │     │
   ├─────┼─────┼─────┼─────┤
   │     │     │     │     │
   └─────┴─────┴─────┴─────┘
```

8.

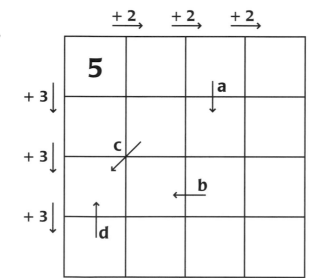

10.

```
      ×5
   ┌─────┬─────┬─────┐
×2 │     │     │     │
   ├─────┼─────┼─────┤
   │     │ 700 │     │
   ├─────┼─────┼─────┤
   │     │     │     │
   └─────┴─────┴─────┘
```

Name _____

Use with *Expressions and Formulas,* page 38.

Jim MacIntosh Total Repairs
147 Franklin Rd., Wakeshire

Customer:_____

Labor _____hours at $37/hour . . . .$_____

Travel costs: . . . .$ 25.00

Total cost per worker . . . .$_____

Total bill = total cost per worker × 3 . . . . . .$_____

(3 workers)

Jim MacIntosh Total Repairs
147 Franklin Rd., Wakeshire

Customer:_____

Labor _____hours at $37/hour . . . .$_____

Travel costs: . . . .$ 25.00

Total cost per worker . . . .$_____

Total bill = total cost per worker × 3 . . . . . .$_____

(3 workers)

Jim MacIntosh Total Repairs
147 Franklin Rd., Wakeshire

Customer:_____

Labor _____hours at $37/hour . . . .$_____

Travel costs: . . . .$ 25.00

Total cost per worker . . . .$_____

Total bill = total cost per worker × 3 . . . . . .$_____

(3 workers)

7.

0.8   10   0.6

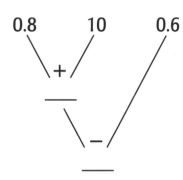

16   1.2   4

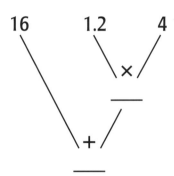

5   7   3   0.6

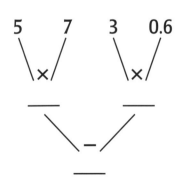

GET YOUR PROGRAMS HERE!

Use additional paper as needed.

Jim is selling programs at a Cubs' game.
He is paid $5.00 per game. In addition,
he gets $0.10 for each program he sells.

1. a. What are his total earnings
if he sells 75 programs at one game?

b. What are his total earnings if he sells
200 programs at one game?

2. If you know the number of programs Jim sold
one day, write a formula to calculate his total earnings for that game.

3. At the first game of the season he made $18.60. How many programs did he sell?

4. Jim's friend sold 350 programs. Make an arithmetic tree using the numbers shown
below to find how much his friend earned.

$5.00   350 $0.10

Name_____ Date_____

Use additional paper as needed.

To be able to fight fires in a large National Park, several airfields are built. From each airfield, airplanes carrying loads of water can take off to fight a fire. It is important that the planes reach the fire within a reasonable amount of time.

There is a formula that describes the average time it takes to get to a fire:

t = d ÷ s + h

t = total time
d = distance in miles
s = speed in miles per hour
h = handling time in hours (the time between a call for help and the takeoff)

1. a. Write the formula in arrow language.

b. Explain the formula.

2. Use the back of this page to calculate the total time to get to a fire if:
  • it is 100 miles from the airfield;
  • the plane flies at an average speed of 50 miles per hour;
  • the handling time is 0.25 hour.

Use additional paper as needed.

3. What is the maximum distance you can have from any point in the park to an airfield if you want the maximum total time needed to arrive at the scene to be 3 hours and 45 minutes? The handling time is always 0.25 hour and the average speed of the plane is 50 mph.

Last week a plane had to go to a fire only 150 miles from the airfield. However, it took 10 hours to get there.

4. a. Does this mean that the formula has to be changed?

b. Describe the factors that can influence the time needed and how they can be avoided.

c. If the total time has to be shortened, where could one save time, the handling time or the average speed?

EXPRESSIONS

Use additional paper as needed.

Compare the following mathematical expressions:

a. $4 \xrightarrow{\times 5} \ldots \xrightarrow{+ 5} \ldots \xrightarrow{\times 4} \ldots$

b.

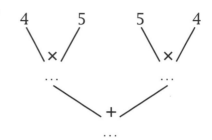

c. $4 \times 5 + 5 \times 4 = \ldots$
(as the calculator you used earlier would compute it)

1. Do these expressions give the same or different results? Explain why.

2. Rewrite part **c** as an expression using parentheses.

Section A. Arrow Language

1. a. October 22: $593.89

October 29: $628.89

b. October 22: $210.24 $\xrightarrow{+\ \$523.65}$ $733.89 $\xrightarrow{-\ \$140}$ $593.89

October 29: $593.89 $\xrightarrow{+\ \$75}$ $668.89 $\xrightarrow{-\ \$40}$ $628.89

c. October 29

2. a. 23

b. 12.8

c. 6080

Section B. Smart Calculations

1. a. $9.59 $\xrightarrow{+\ \$0.01}$ $9.60 $\xrightarrow{+\ \$0.05}$ $9.65 $\xrightarrow{+\ \$0.10}$ $9.75 $\xrightarrow{+\ \$0.25}$ $10.00 $\xrightarrow{+\ \$10.00}$ $20.00

$9.59 $\xrightarrow{+\ \$10.41}$ $20.00

b. $2.26 $\xrightarrow{+\ \$0.04}$ $2.30 $\xrightarrow{+\ \$0.20}$ $2.50 $\xrightarrow{+\ \$0.50}$ $3.00 $\xrightarrow{+\ \$2.00}$ $5.00

$2.26 $\xrightarrow{+\ \$2.74}$ $5.00

c. $15.64 $\xrightarrow{+\ \$0.01}$ $15.65 $\xrightarrow{+\ \$0.10}$ $15.75 $\xrightarrow{+\ \$0.25}$ $16.00

$15.64 $\xrightarrow{+\ \$0.36}$ $16.00

2. a. 750 $\xrightarrow{+\ 75}$ _____

b. 63 $\xrightarrow{+\ 47}$ _____

c. 439 $\xrightarrow{-\ 19}$ _____

3. Answers will vary. One possibility is shown below for each part.

a. 74 $\xrightarrow{+\ 6}$ _____ $\xrightarrow{+\ 60}$ _____

It is easier to add 66 when it is split up into two parts. This way I can first add the ones part and then the tens part.

b. 231 $\xrightarrow{-\ 60}$ _____ $\xrightarrow{+\ 2}$ _____

It is easier to subtract 58 by first subtracting 60 (a multiple of 10) and then adding two back on.

c. 459 $\xrightarrow{+\ 21}$ _____ $\xrightarrow{+\ 6}$ _____

I added 459 + 21 first because 480 is easy to work with. I then added six on because I had to add on a total of 27.

Student Page

TRY THIS!

Section A. Arrow Language

1. Here is a record for Mr. Kamarov's bank account.

Date	Deposit	Withdrawal	Total
10/15			$210.24
10/22	$523.65	$140.00	
10/29	$75.00	$40.00	

a. Find the totals for October 22 and October 29.

b. Write arrow strings to show how you found the totals in part **a**.

c. When does Mr. Kamarov first have at least $600 in his account?

2. Find the results for the following arrow strings:

a. 15 $\xrightarrow{-\ 3}$ _____ $\xrightarrow{+\ 11}$ _____

b. 3.7 $\xrightarrow{+\ 1.9}$ _____ $\xrightarrow{+\ 8.8}$ _____ $\xrightarrow{-\ 1.6}$ _____

c. 3,000 $\xrightarrow{-\ 1,520}$ _____ $\xrightarrow{-\ 600}$ _____ $\xrightarrow{+\ 5,200}$ _____

Section B. Smart Calculations

1. Below are some shopping problems. For each, write an arrow string to show the change that the cashier gives to the customer. Be sure to use the small-coins-and-bills-first method. Then write another arrow string that has only one arrow to show the total change.

a. A customer gives $20.00 for a $9.59 purchase.

b. A customer gives $5.00 for a $2.26 purchase.

c. A customer gives $16.00 for a $15.64 purchase.

2. Rewrite the following arrow strings so that each has only one arrow:

a. 750 $\xrightarrow{+\ 35}$ _____ $\xrightarrow{+\ 40}$ _____

b. 63 $\xrightarrow{-\ 3}$ _____ $\xrightarrow{+\ 50}$ _____

c. 439 $\xrightarrow{+\ 1}$ _____ $\xrightarrow{-\ 20}$ _____

3. Rewrite each of the following arrow strings with a new string that will make the computation easier. Explain either why your new string makes the computation easier or why this is not possible.

a. 74 $\xrightarrow{+\ 66}$ _____

b. 231 $\xrightarrow{-\ 58}$ _____

c. 459 $\xrightarrow{+\ 27}$ _____

54   Britannica Mathematics System

Section C. Formulas

1. String **c** gives the correct cost because the number of hours are multiplied by two, the rate per hour, and the monthly cost is added on separately.

2. **a.** $25

   b. $55

   c. $28

3. number of hours $\xrightarrow{\times\ \$3}$ _____ $\xrightarrow{+\ \$10}$ total cost

4. Clarinda should use *Tech Net*. With *Tech Net*, her cost is $35 a month for 10 hours, whereas with *Online Time* her cost is $40 for 10 hours.

5. two pots: 24 centimeters
   three pots: 28 centimeters

6. Answers will vary. Sample solution:

   number of pots $\xrightarrow{\times\ 4\ cm}$ _____ $\xrightarrow{+\ 16\ cm}$ stack height

7. Seven pots. Explanations will vary. Sample explanations:

   Only seven pots will fit in this stack because if you use the arrow strings to multiply seven by four and add on 16 you get 44 centimeters. The stack must be less than 45 centimeters, so only seven pots will fit.

   I reversed the arrow string and subtracted 16 from 45 centimeters (29) and then divided by four (7.25). Since you can't stack 7.25 pots, the closest amount that will fit is seven pots.

8. **a.** same results

   b. different results

   c. same results

Student Pages

Try This!

Section C. Formulas

Clarinda has a personal computer at home, and she subscribes to Tech Net for Internet access. Tech Net charges $15 per month for access plus $2 per hour of usage. For example, if Clarinda is connected to the Internet for a total of three hours one month, she pays $15 plus three times $2, or $21, for the month.

1. Which of these strings gives the correct cost for Internet service through Tech Net? Explain your answer.
   a. $15 $\xrightarrow{+\ \$2}$ _____ $\xrightarrow{\times\ number\ of\ hours}$ total cost
   b. number of hours $\xrightarrow{+\ \$15}$ _____ $\xrightarrow{\times\ \$2}$ total cost
   c. number of hours $\xrightarrow{\times\ \$2}$ _____ $\xrightarrow{+\ \$15}$ total cost

2. How much does it cost Clarinda for the following amounts of monthly usage:
   a. 5 hours
   b. 20 hours
   c. $6\frac{1}{2}$ hours

Another company, Online Time, charges only $10 per month, but $3 per hour.

3. Write an arrow string that can be used to find the cost of Internet access through Online Time.

4. If Clarinda uses the Internet approximately 10 hours a month, which company should she use—Tech Net or Online Time?

Mathematics in Context • Expressions and Formulas   55

Expressions and Formulas

Carlos works at a plant store that sells flower pots. One type of flower pot has a rim height of 4 centimeters and a hold height of 16 centimeters.

5. How tall is a stack of two of these pots? three of these pots?

6. Write a formula using arrow language that can be used to find the height of a stack if you know the number of pots.

7. Carlos has to stack these pots on a shelf that is 45 centimeters high. How many can be placed in a stack this high? Explain your answer.

8. Compare the following pairs of arrow strings and decide whether they provide the same or different results:
   a. input $\xrightarrow{\times\ 8}$ _____ $\xrightarrow{\div\ 2}$ output
   input $\xrightarrow{\div\ 2}$ _____ $\xrightarrow{\times\ 8}$ output
   b. input $\xrightarrow{+\ 5}$ _____ $\xrightarrow{\times\ 3}$ output
   input $\xrightarrow{\times\ 3}$ _____ $\xrightarrow{+\ 5}$ output
   c. input $\xrightarrow{\div\ 2}$ _____ $\xrightarrow{+\ 1}$ _____ $\xrightarrow{+\ 6}$ output
   input $\xrightarrow{\div\ 2}$ _____ $\xrightarrow{+\ 6}$ _____ $\xrightarrow{+\ 1}$ output

56   Britannica Mathematics System

Section D. Reverse Operations

1. **a.** $1.50

   b. $13.50

   c. $6

   d. $33.75

2. Number of U.S. dollars $\xrightarrow{\div 3}$ _____ $\xrightarrow{\times 4}$ number of Canadian dollars

3. **a.** input $\xleftarrow{+ 1}$ _____ $\xleftarrow{\div 2.5}$ _____ $\xleftarrow{- 4}$ output

   b. input $\xleftarrow{- 6}$ _____ $\xleftarrow{+ 2}$ _____ $\xleftarrow{\times 5}$ output

4. **a.** 14

   b. 8

Section E. Tables

1. **a.**

   $\xrightarrow{+5}$
   $\downarrow +1$

2	7	12	17
3	8	13	18
4	9	14	19
5	10	15	20

   b.

   $\xrightarrow{\times 4}$
   $\downarrow \times 3$

2	8	32	128
6	24	96	384
18	72	288	1,152
54	216	864	3,456

2. **a.** × 3

   b. ÷ 2

   c. × 6

   d. × 4

Student Page

Try This!

Section D. Reverse Operations

Ravi lives in Bellingham, Washington. He travels to Vancouver, Canada, quite frequently. When Ravi is in Canada, he uses the following rule to estimate prices in U.S. dollars:

number of Canadian dollars $\xrightarrow{+4}$ _____ $\xrightarrow{\times 3}$ number of U.S. dollars

1. Using Ravi's formula, estimate U.S. prices for the following Canadian prices:
   a. a hamburger for $2 Canadian
   b. a T-shirt for $18 Canadian
   c. a movie for $8 Canadian
   d. a pair of shoes for $45 Canadian

2. Write a formula that Ravi can use for converting U.S. dollars to Canadian dollars.

3. Write the reverse string for each of the following strings:
   a. input $\xrightarrow{-1}$ _____ $\xrightarrow{\times 2.5}$ _____ $\xrightarrow{+4}$ output
   b. input $\xrightarrow{+6}$ _____ $\xrightarrow{-2}$ _____ $\xrightarrow{+5}$ output

4. Find the input for each of the following strings:
   a. input $\xrightarrow{+10}$ _____ $\xrightarrow{+2}$ _____ $\xrightarrow{-3}$ 9
   b. input $\xrightarrow{\times 4}$ _____ $\xrightarrow{-5}$ _____ $\xrightarrow{+3}$ _____ $\xrightarrow{+1}$ 10

Section E. Tables

1. In your notebook, copy and complete the following tables.

   a.

   b.

2. Which operations fit with the arrows marked **a** through **d** inside the following table?

Mathematics in Context • Expressions and Formulas

57

3. a. Answers will vary. Sample responses:

For women's shoe sizes, as you go down each column, the size increases by $\frac{1}{2}$. An equivalent U.K. shoe size is $1\frac{1}{2}$ sizes smaller than the U.S. shoe size. An equivalent European shoe size is 31 sizes larger than the U.S. shoe size.

For U.S. and U.K. men's shoe sizes, as you go down each column, the size increases by 1. An equivalent U.K. shoe size is 1 size smaller than the U.S. shoe size.

b. horizontal string: U.S. shoe size $\xrightarrow{-1\frac{1}{2}}$ U.K. shoe size $\xrightarrow{+32\frac{1}{2}}$ European shoe size

vertical string:

U.S. shoe size	U.K. shoe size	European shoe size
$\downarrow +\frac{1}{2}$	$\downarrow +\frac{1}{2}$	$\downarrow +\frac{1}{2}$
next largest U.S. shoe size	next largest U.K. shoe size	next largest European shoe size

c. It is not possible to write an arrow string for men's European shoe sizes because the pattern is not regular. The following strings would work for U.S. and U.K. men's shoe sizes:

horizontal string: U.S. shoe size $\xrightarrow{-1}$ U.K. shoe size

vertical string:

U.S. shoe size	U.K. shoe size
$\downarrow +1$	$\downarrow +1$
next largest U.S. shoe size	next largest U.K. shoe size

Section F. Order of Operations

1. a.

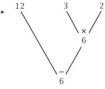

b.

c.

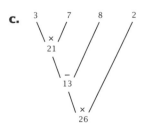

Student Page

Expressions and Formulas

3. Look carefully at the following tables for women's and men's U.S., U.K., and European shoe sizes.

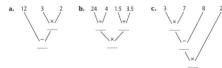

Women's Shoe Sizes

U.S.	U.K.	Europe
5	3½	36
5½	4	36½
6	4½	37
6½	5	37½
7	5½	38
7½	6	38½
8	6½	39
8½	7	39½
9	7½	40

Men's Shoe Sizes

U.S.	U.K.	Europe
6	5	38
7	6	39½
8	7	41
9	8	42
10	9	43
11	10	44½
12	11	46
13	12	47
14	13	48

Source: Data from *How Many, How Long, How Far, How Much*
(The Stonesong Press, Inc., 1996).

a. Make a list of all of the regularities you can find in the tables above.

b. If possible, write a horizontal and a vertical arrow string that could be used to generate the table for women's shoe sizes. If this is not possible, explain why.

c. Do the same as in part **b** for men's shoe sizes.

Section F. Order of Operations

1. In your notebook, copy and complete the following arithmetic trees:

a. 12 3 2   **b.** 24 4 1.5 3.5   **c.** 3 7 8 2

2. Draw an arithmetic tree and find the answer for the following calculations:
   a. $10 + 1.5 \times 6$ **b.** $(10 + 1.5) \times 6$ **c.** $15 \div (2 \times 2 + 1)$

3. Suzanne had to go to the veterinarian because her cat needed dental surgery. (Her cat never brushed its teeth!) Before the surgery, the veterinarian gave Suzanne the following estimate for the cost: $55 for anesthesia, $30 total for teeth cleaning, $18 per tooth pulled, $75 per hour of surgery, plus the cost of medicine. Make an arithmetic tree that could be used to find the total cost of Suzanne's bill from the veterinarian. Use words in your arithmetic tree where necessary.

58

Britannica Mathematics System

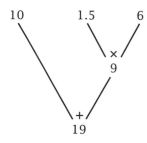

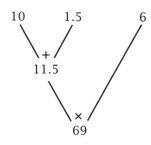

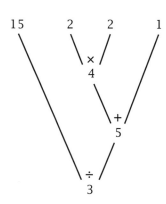

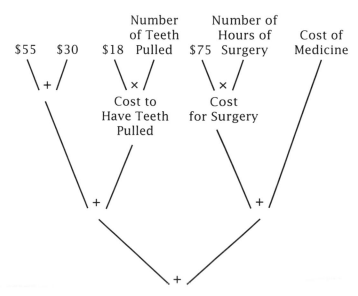

CREDITS

Cover

Design by Ralph Paquet/Encyclopædia Britannica, Inc.

Collage by Koorosh Jamalpur/KJ Graphics.

Title Page

Brent Cardillo/Encyclopædia Britannica, Inc.

Illustrations

6, 8, 10 Phil Geib/Encyclopædia Britannica, Inc.; **12** Paul Tucker/Encyclopædia Britannica, Inc.; **14, 16** Phil Geib/Encyclopædia Britannica, Inc.; **20, 24** Paul Tucker/Encyclopædia Britannica, Inc.; **26** Brent Cardillo/Encyclopædia Britannica, Inc.; **30, 32** Phil Geib/Encyclopædia Britannica, Inc.; **38, 40** Paul Tucker/Encyclopædia Britannica, Inc.; **50, 52, 54, 56, 64, 66, 68, 70, 72** Phil Geib/Encyclopædia Britannica, Inc.; **74** Paul Tucker/ Encyclopædia Britannica, Inc.; **78** Brent Cardillo/Encyclopædia Britannica, Inc.; **80** (top) Phil Geib/Encyclopædia Britannica, Inc.; **80** (bottom) Paul Tucker and Tom Zielinski/Encyclopædia Britannica, Inc.; **82, 86, 88, 90** Phil Geib/Encyclopædia Britannica, Inc.; **96** JeromeGordon/Encyclopædia Britannica, Inc.; **104** Phil Geib/Encyclopædia Britannica, Inc.; **112, 114** Paul Tucker/Encyclopædia Britannica, Inc.; **124** Jerome Gordon/Encyclopædia Britannica, Inc.; **126** Brent Cardillo/Encyclopædia Britannica, Inc.

Photographs

42 © Ezz Westphal/Encyclopædia Britannica, Inc.